Inkjet Based 3D
Additive Manufacturing of Metals

Mojtaba Salehi, Manoj Gupta
Saeed Maleksaeedi, Nai Mui Ling Sharon

Additive Manufacturing (AM) is a highly promising rapid manufacturing process. Based on incremental layer-upon-layer deposits, three dimensional components of high geometrical complexity can be produced; applications ranging from aerospace and automotive to biomedical industries. Laser, electron beam and wire-based techniques are reviewed.

Particular emphasis is placed on 3D inkjet printing of metals, which is reviewed here in great depth and for the first time. This is an ambient temperature technology which offers some unique advantages for printing metals and alloys, as well as composite and functionally graded materials. Material selection guidelines are presented and the various deposition techniques and post-printing treatments are discussed; together with the resulting properties of the printed components: Density, shrinkage, resolution and surface roughness, porosity-related and mechanical properties, as well as biological properties

The various metal printing techniques are compared with each other and case studies are referred to.

Inkjet Based 3D Additive Manufacturing of Metals

Mojtaba Salehi
Manoj Gupta

Department of Mechanical Engineering,
National University of Singapore,
Singapore

Saeed Maleksaeedi
Nai Mui Ling Sharon

3D Additive Manufacturing Team,
Forming Technology Group,
Singapore Institute of Manufacturing,
Technology (SIMTech),
Singapore

Published by **Materials Research Forum LLC**
Millersville, PA 17551, USA

Published as part of the book series
Materials Research Foundations
Volume 20 (2018)
ISSN 2471-8890 (Print)
ISSN 2471-8904 (Online)

Print ISBN 978-1-945291-44-9
ePDF ISBN 978-1-945291-45-6

This book contains information obtained from authentic and highly regarded sources. Reasonable efforts have been made to publish reliable data and information, but the author and publisher cannot assume responsibility for the validity of all materials or the consequences of their use. The authors and publishers have attempted to trace the copyright holders of all material reproduced in this publication and apologize to copyright holders if permission to publish in this form has not been obtained. If any copyright material has not been acknowledged please write and let us know so we may rectify in any future reprint.

Distributed worldwide by

Materials Research Forum LLC
105 Springdale Lane
Millersville, PA 17551
USA
http://www.mrforum.com/

Manufactured in the United States of America
10 9 8 7 6 5 4 3 2 1

Preface

Additive manufacturing (AM) is a promising rapid manufacturing process with unrivalled advantages. Unlike conventional manufacturing methods, AM techniques are versatile and provide unsurpassed design freedom and short lead time enabling to produce even a small volume of critical parts since neither tooling nor special fixtures are required. AM offers the most value-added benefits to fabricate near-net shapes and geometrically complex components made of conventional and high-value materials. As a result, serious efforts are made by researchers worldwide to master different AM techniques so that it can be scaled up for a wide spectrum of industrial sectors. AM techniques are principally based on the fundamental concept of depositing incremental layer-upon- layer to realize a three-dimensional object. A desired CAD model is the starting point followed by slicing the CAD data into layers with specific thicknesses. For the powder bed AM methods, a roller spread feedstock in the predetermined fashion with the predefined layer thickness. Subsequently, the deposited feedstock binds/melts selectively in accordance with the sliced CAD data of each layer. Repetition of such layer by layer steps eventually builds the entire physical component.

Multiple engineering sectors such as aerospace, automotive and biomedical sectors are targeting to integrate AM techniques in their manufacturing section as a cost-effective option for production of small batches and the possibility of parts' customization. In this respect, AM systems based on either laser or electron beam melting have been recently favoured and extensively investigated. However, high capital expenditure of equipment including their maintenance cost hinder wider applications of these techniques. Among the other AM systems for metallic materials, inkjet three-dimensional printing (3DP) offers some unique advantages. For instance, 3DP is an ambient temperature technology for metals processing. To date no attempt has been made to integrate the state of knowledge available in the open literature for inkjet based 3DP of metals. Accordingly, the principal objective of this book is to provide readers an insight into the existing AM methodologies for metal 3DP and how to judiciously select materials and 3DP parameters. Particular emphasis is placed on inkjet printing metallic systems and their properties as performance indicators to illustrate the opportunities for 3DP of metallic parts.

The authors feel strongly that this book will be of interest to a variety of readers ranging from students and teachers to engineers and researchers who are particularly interested in 3DP of metals.

Acknowledgments

We would like to express our heartiest thanks to our families for their unflagging patience and cheerful support. Our appreciation also goes to our colleagues, friends and students for their continued inspiration. Finally, we would like to take this opportunity to thank the people who have contributed to and assisted with the publication of this book.

Mojtaba Salehi
Manoj Gupta

Department of Mechanical Engineering,
National University of Singapore,
Singapore

Saeed Maleksaeedi
Nai Mui Ling Sharon

3D Additive Manufacturing Team,
Forming Technology Group,
Singapore Institute of Manufacturing,
Technology (SIMTech),
Singapore

Table of Contents

CHAPTER 1

Overview of Additive Manufacturing Processes for Metals

Abstract

This chapter introduces classifications of the techniques used for metal additive manufacturing. Further, it provides a basic understanding of the working principals of powder bed and directed energy deposition processes for metal additive manufacturing, including laser and electron beams based techniques. Serving as an additive manufacturing process for all three types of materials, inkjet three-dimensional printing (3DP) is introduced with the focus on metals. Its advantages and disadvantages are thoroughly discussed.

Keywords

Additive Manufacturing, Electron Beam, Laser Beam, Inkjet Three-Dimensional Printing (3DP), Metals

1.1 Introduction to Additive Manufacturing

Additive manufacturing (AM), also regarded as rapid prototyping (RP) or solid free-form fabrication (SFF), is a process of joining materials to make objects from 3D model data, usually layer upon layer, as opposed to subtractive manufacturing methodologies [1]. AM techniques have generated a growing interest in recent years among researchers to build parts with great geometrical complexities which otherwise would be difficult to fabricate using conventional manufacturing processes (i.e. subtractive and formative). Neither tooling nor special fixtures are required in AM resulting in significantly shortened lead time. Other benefits of AM processes include:

a) ability to fabricate near net shape parts,

b) cost-effective for small batch production, and

c) possibility of parts customization.

Automotive [2], aerospace [3, 4], and biomedical [5, 6] are identified as promising sectors for AM applications. In general, all AM processes begin with a computer-aided-design (CAD) file which is converted to the Standard Triangle (Tessellation) Language (STL) file format or additive manufacturing format (AMF). This file is transferred to an

AM machine and manipulated in accordance with the AM technique that is to be used. Upon finishing layer-by-layer build cycle, the fabricated part is removed from the AM machine. Some post-processing may be required for the fabricated part to deliver its full functionality [7, 8].

Stereolithography (SLA/SL) emerged as first AM method in which a liquid photosensitive polymer solidifies with the use of an ultraviolet (UV) laser beam [9]. SLA was patented in 1986, and since then other AM systems have been introduced.

Metals have been traditionally the backbone of human civilization and its development due to their ability to exhibit diverse combination of properties. Accordingly, they are used in numerous engineering and biomedical applications. Given the great benefits of AM to open new opportunities for industries, it is not surprising that AM for metals is expanding at a rapid pace through a wide spectrum of industries.

1.2 Metal Additive Manufacturing

AM techniques for metals can broadly be categorized in terms of how raw materials and heat sources interact. AM processes are classified as: (i) material deposition methods or directed energy deposition, in which raw materials in the form of a wire or powder are fused by means of thermal energy sources such as laser or electron beam and deposited layer by layer and (ii) powder bed methods or powder bed fusion, wherein raw materials in the form of a thin layer of powder are spread and either selectively fused by thermal energy sources or bonded by means of adhesive materials [1, 8, 10-12]. Another way to classify AM techniques is based on the number of steps required to fabricate parts [12]. Accordingly, AM techniques are divided into two categories: (i) single-step processes and (ii) multi-step processes. In the single-step processes, an intended part is obtained in a single operation, while the multi-step processes involve at least two steps, including (a) creation of the geometric shape of a part and (b) consolidation of the created part to obtain the anticipated properties. Fig. 1.1 presents an overview of metal AM processes. Bonding methods of each technique, along with an example of machine manufacturer for some techniques are also provided in Fig. 1.1.

Recently, laser and electron beam based AM processes have become the most prevailing AM for metals. Many attempts have been made to summarize current state-of-the-art in the obtained properties of a certain metal using a specific AM technique. The list includes but is not limited to selective laser melting/sintering of aluminum alloys [13, 14], magnesium [15], titanium [16, 17], nickel-based superalloy [18] or electron beam melting of Ti [19], iron based metals, and cobalt-base alloys [20, 21]. Thus, familiarizing with the working principles of these processes is essential. Fundamentals of these processes are summarized in the forthcoming sections.

1.2.1 Laser Based AM Processes

As shown in Fig. 1.1, three laser based AM techniques are laser melting, laser sintering, and laser powder deposition [22]. It should be pointed out these processes may be addressed by other names, according to their trademarks. For example, the laser powder deposition technique has several trademarks among which Laser Engineered Net Shaping (LENS) and Direct Metal Deposition (DMD) are the most famous ones. Despite distinctive names and slight differences in DMD and LENS, both systems share a similar concept of coaxial powder feeding with synchronous laser scanning.

1.2.1.1   Laser Melting

Principals of laser melting, popularly known as selective laser melting (SLM), is schematically illustrated in Fig. 1.2. Before importing the CAD file into an SLM machine, the file is processed by a software to provide required support structures which are eventually appended to the STL file. The fabrication steps start with fixing and levelling of a substrate plate on the build platform. The part will be built on this substrate which can be pre-heated in some machines. The build chamber is then filled with either argon or nitrogen gas to minimize undesirable reactions as well as to provide cooling and heat conduction [23]. Afterwards, a layer of powder with thickness between 20 to 100 μm is spread on the substrate plate and selectively melted by means of a focused laser beam (e.g., 100 μm in diameter) in accordance with the 2D cross-sectional area. By lowering the build platform by the thickness of a layer, the subsequent layer of powder is spread on the top of the preceding layer and fused by the laser beam. Eventually, the 3D part is built by repeating the spreading and laser scanning for consecutive layers of powder in a layer by layer manner. When the build cycle is complete, the part is removed from the loose supporting powder and cut from the substrate plate. Post-processing steps such as hot isostatic pressing (HIP) and heat treatment might be carried out to release residual stresses and further enhance the properties of the fabricated parts [8, 24].

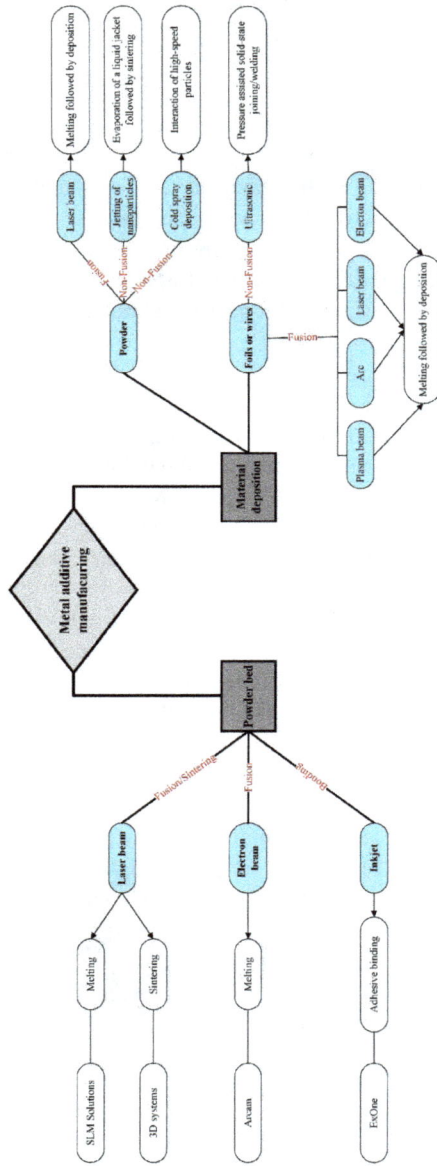

Figure 1.1 Summary of AM processes for metals, along with a few examples of machine manufacturers [8, 11].

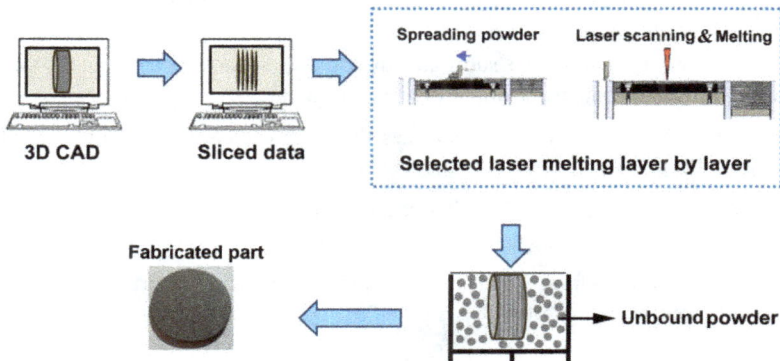

Figure 1.2 A schematic representation of an SLM process (adapted from [25]).

1.2.1.2   Laser Sintering

Laser sintering is also known as selective laser sintering. It was the first commercialized powder bed AM technique patented in the late 1980s [26]. This technique shares similar procedures and processing apparatus with laser melting as described above. However, the laser melting is based on a melting-solidification concept, while in laser sintering, the laser beam only heats the powder bed enabling the loose powder particles to bond with each other. This results into sintering into a porous or a solid mass in the absence of complete liquefaction. Essentially, two types of powder beds can be used with laser sintering machines: (i) a mixture of metal powder and polymer in which the polymeric material is melted to stick the metal powder together and (ii) the sole metal powder in the absence of polymer materials is sintered to bind metal particles. Comprehensive comparisons (in terms of mechanical properties, surface quality, dimensional accuracy, building speed, operation cost, and material issues) were made between these two approaches elsewhere [27].

1.2.1.3   Laser Powder Deposition

Schematic of laser powder deposition (LPD) technology, also known as LENS or DMD techniques, is depicted in Fig. 1.3. Materials in powder form are delivered by means of an inert gas through specially designed nozzles directed towards a substrate. Concurrently, the powder is melted by a high energy laser beam, which is focused in a close contiguity of the substrate, and deposited due to the blowing gas. With moving the build platform and/or the nozzles, desired cross-sectional features are additively deposited at the right place in a layer by layer manner to produce a 3D structure. LPD can deliver powder in a wide spectrum of angles from 0 to 180° with respect to the substrate [28]. Finally, the fabricated part is removed from the substrate. In addition to laser beam sources and optical systems, the main components of a LPD machine include:

i. nozzles which are commonly coaxial ones that enable uniform deposition in all directions [29],

ii. feedback control system which gather real-time information of the melt pool to adjust processing parameters, and

iii. five/six-axis build platform which uses computer aided manufacturing (CAM) to selectively control the materials deposition [22].

LPD technology has a strong potential in remanufacturing and repair of worn structures such as turbine blades and shafts as well as in cladding and hardfacing of molds and dies, along with serving as an AM machine [30-32]. Moreover, the manner of powder supply and deposition makes it possible to develop new materials and optimally designed structures by tailoring microstructure and composition within the part being fabricated using LPD [22]. The downside of LPD is associated with low powder deposition efficiency (~ 30%) even when powder recycling is being practiced [28].

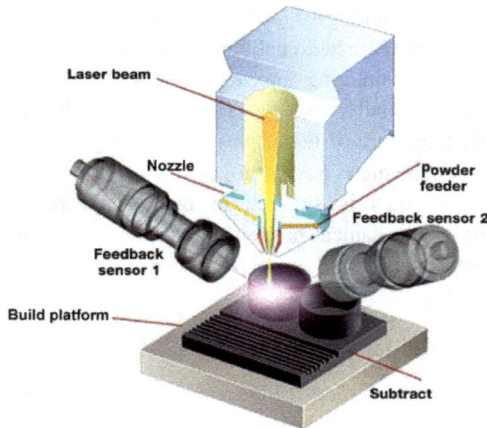

Figure 1.3 A schematic illustration of laser powder deposition process [33].

1.2.2 Electron Beam Melting

In electron beam melting (EBM) metal powder is melted using the heat generated by interaction of an electron beam with the powder under a high vacuum condition. A basic vacuum of $<10^{-4}$ Torr is provided within the build chamber using a vacuum system attached to the EBM systems. However, a helium gas bleeding makes it possible to increase the pressure of the build area to the order of 10^{-2} Torr in order to improve heat conduction and provide cooling [34, 35]. Commercially available EBM machines are patented by Arcam, Sweden. Fig. 1.4a displays a schematic architecture of an Arcam's EBM system wherein the electron gun assembly generates electron beams which are accelerated and focused by the electron optical system including a couple of electromagnetic lenses. A thin layer of powder (<100 µm) is deposited onto the substrate plate by raking gravity fed powder which flows from the powder hopper. As illustrated in Fig. 1.4b, each layer of powder is initially preheated to 0.4-0.8 T_m (where T_m is the melting temperature of the powder) in the course of multi-passes rapid scanning with a large focal beam spot [34, 35]. Subsequently, the preheated layer is selectively melted in accordance with the STL data using the electron beams with reduced current and at slower scan rate. Afterwards, the build platform moves down by one layer thickness and then a new layer of the powder is spread, preheated and selectively melted (Fig. 1.4b). These steps are repeated until the entire part is fabricated. After the build cycle is completed, the build chamber is left to cool to ambient temperature which is followed by removal of loose powder and detachment of the part from the substrate.

Higher energy efficiency, shorter build cycles, and lesser residual stress within the fabricated components are some distinct advantages of EBM over SLM [36-38]. Several comparative investigations into microstructural characteristics, biocompatibility, physical and mechanical properties of EBM and SLM fabricated parts have been made to explore inherent properties of each technique [19, 39-43].

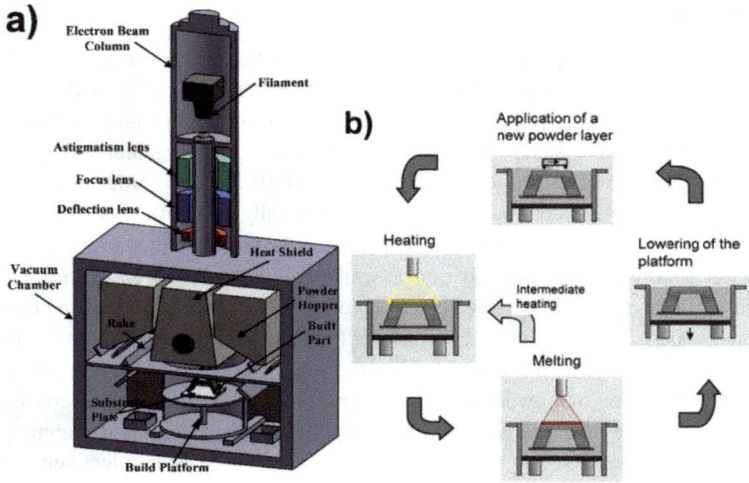

Figure 1.4 a) A schematic architecture of an Arcam EBM machine [44] and b) processing steps to build one layer using a EBM machine [20].

1.2.3 Wire-based AM Techniques

Instead of using metal powder, fine metal wires are melted and deposited in a layer-by-layer fashion on a substrate plate to build up components as depicted in Fig. 1.5. According to the energy sources used, wire-based AM techniques can be categorized into three classes: (i) arc or plasma based, (ii) laser based, and (iii) electron beam based methods [45]. Moreover, these techniques can be used in conjunction with conventional manufacturing methods (e.g., milling) as a hybrid method [45, 46]. The major advantages of wire based techniques over powder based ones include:

   i. higher quality of raw materials due to less contaminations in the form of wires,

  ii. reduced materials cost,

 iii. higher deposition rate,

  iv. much higher deposition efficiency, and

   v. the possibility of fabricating components with larger size [47-50].

However, as a result of high heat input, high distortions and residual stress are associated with wire based processes [51]. In contrast to wide deposition angles in powder deposition techniques, the deposition angle with respect to the substrate for wire feeding is quite restricted (e.g., 10–75°) [28, 52]. Furthermore, poorer surface finish and dimensional accuracy compared to the powder based counterparts further hinder the wider applications of wire based techniques [51]. Nevertheless, wire based AM technologies can be taken into accounts as alternative candidates for manufacturing components with either moderate complexity and larger features or median to large size, offering cost-competitiveness, greater materials quality, and energy efficiency compared to the powder based processes [51].

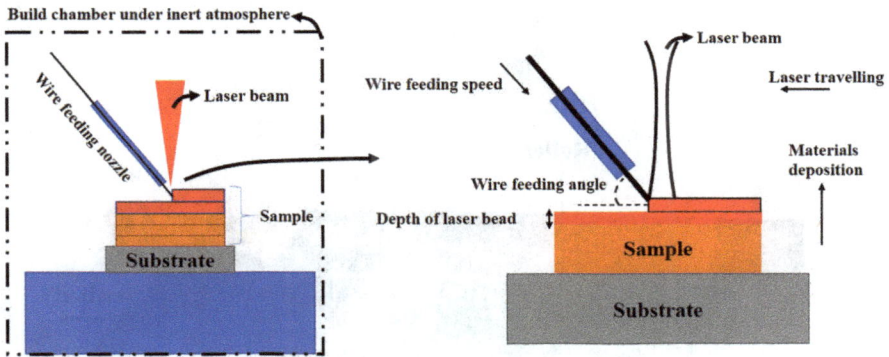

Figure 1.5 A schematic architecture of wire based laser melting deposition apparatus (adapted from [49]).

1.2.4 Inkjet three-dimensional Printing (3DP)

Inkjet 3DP is a powder-based AM method invented at the Massachusetts Institute of Technology [53]. Before printing starts, a pre-designed 3D model data is converted into a pile of two-dimensional (2D) slices with certain thickness, representing the 3D model. The final part is fabricated by the successive printing of these 2D layers. As depicted in Fig. 1.6, each 2D layer is created by moving up the feedstock platform in conjunction with the build platform which moves down one layer thickness of the slice in height. Next, the feedstock is conveyed to the build platform from the feedstock platform by a roller. By means of the print head rastering across the powder bed, liquid droplets are then selectively deposited into the locations delineated by a sliced 2D object profile, causing the feedstock to stick together with the previously deposited layer. These steps are repeated layer by layer until the end of the build cycle. Finally, the as-built part is removed from the powder bed followed by post-printing steps which include burning out the binder materials and sintering.

Final metallic parts can entirely be fabricated by inkjet 3DP which regards as direct inkjet 3DP. On the other hand, indirect 3DP process involves inkjet 3DP of a casting mold, infiltrating/pouring a molten metal into the mold, and removing/dissolving the mold to leave a final part [54-56]. Process flowcharts for direct and indirect metal inkjet 3DP are depicted in Fig. 1.7. Indirect 3DP, in which extensive post-manufacturing processes are required, is not addressed in this book. The scope of this book is devoted to the direct inkjet 3DP of metals.

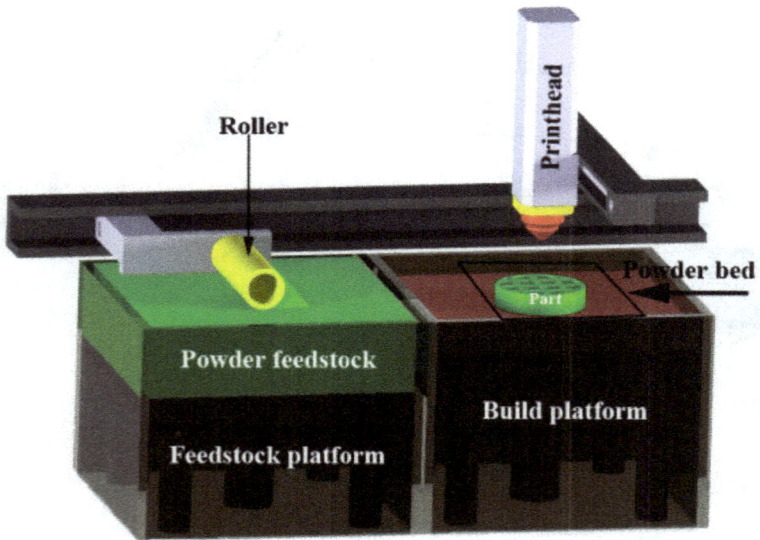

Figure 1.6 A schematic illustration of inkjet 3DP process.

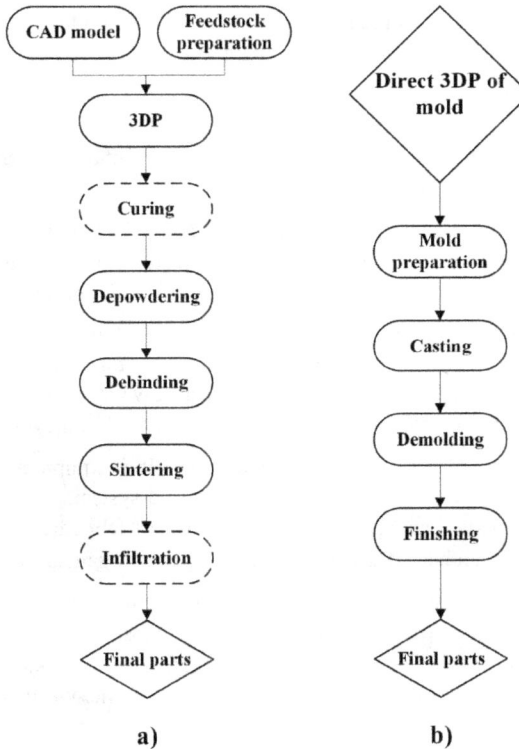

Figure 1.7 Process flowcharts for a) direct and b) indirect inkjet 3DP of metals (steps indicated by dash lines are optional).

1.3   Advantages and Disadvantages of the Inkjet 3DP Method

Some advantages of the inkjet 3DP include:

1) Almost any material, including polymers [57], ceramics [58], and metals are printable. Moreover, it is possible to deliver a mixture of materials as feedstock to produce composites component and functionally graded materials.

2) The build cycle is carried out at or near ambient temperature, avoiding issues associated with phase changes, oxidation, and residual stress.

11

3) The build cycle is performed under an atmospheric condition. In other words, no controlled atmosphere is required during inkjet 3DP.

4) The build envelope can be scaled up more readily compared to the other powder bed methods. For instance, the build envelope can be as large as $4000 \times 2000 \times 1000$ mm^3 [59], offering a huge potential for fabrication in large quantities and sizes [60].

5) Inkjet 3DP is a support-free AM technique in which loose powder within the powder bed provide full supports for internal features, undercuts, and overhangs to be created, as against relative support in powder bed fusion techniques.

6) Inkjet 3DP provides more design freedom over SLM and EBM. In SLM or EBM, the geometry of a part along with its support structures should be designed in a manner that the supports not only provide heat flow paths and mechanical supports for the part but also offer an ease of removal after the build cycle [61, 62].

7) Much lower equipment capital expenditure as well as equipment maintenance cost in comparison with laser or electron beams based systems, resulting in lower cost of fabricated parts. For example, the cost of a part fabricated using inkjet 3DP is generally less than half of the similar part fabricated by means of SLS [60].

8) Minimal feedstock wastage and full recycling of used powder are possible. The support-free nature of the process can save substantial amount of feedstock. Moreover, the powder batch is fully recyclable. For the inkjet 3DP process, the used powder has exactly the same properties as the virgin powder, while in powder bed fusion systems, heat flow to the build platform may alter properties of the used powder which adversely affects repeatability and consistency of the part's properties [63, 64].

9) Inkjet 3DP has a promising potential for series production since the build cycle of a part is largely governed by liquid depositing time. Thus, the build time can be reduced by using multiple print heads [65]. Furthermore, inkjet 3DP indicates acceptable repeatability [66]. For example, the accuracy of the printed parts ranges from ±0.13 to ±1.3 mm, depending on the components' dimensions [67].

10) Less process optimization is required for developing a new material system compared to other powder bed or directed energy deposition techniques because of two reasons. Firstly, the deposited liquid that binds the feedstock within the powder bed can be compatible with a variety of materials [68]. Secondly, well-known thermal processing knowledge of conventional powder-based techniques is applicable to a post-printing step.

11) There is neither distortion during the build cycle nor spatial variation in the properties of fabricated parts, as compared to SLM or EBM fabricated parts [8, 64, 69-71].

The main drawbacks of inkjet 3DP include:

1) Inkjet 3DP is a multi-step process where post-printing steps are required for as-built parts.

2) As-built parts have lower relative density (i.e. the ratio of density of as-build parts to the bulk density of feedstock) than those fabricated by the powder bed fusion or directed energy deposition processes [72].

3) The resolution of an inkjet 3D printed part is not as high as other technologies such as SLA or SLM. For examples, the resolution is 0.5 to 50 μm for SLA, compared to 50 to 100 μm for inkjet 3DP [6].

Despite these weaknesses, inkjet 3DP provides distinct advantages for many applications. In the biomedical sector [73-75], for example, an interconnected micro-porosity scaffold fabricated by inkjet 3DP is able to enhance vascularization and support angiogenesis [76-79], whereas such a micro-porosity cannot readily be fabricated with SLM or EBM techniques [19, 80-84].

Summary

This chapter provides a broad overview of current AM processes for metals. The operation principles of metal AM processes, including laser sintering/melting, laser powder deposition, electron beam melting, wire-based processes, and inkjet three-dimensional printing (3DP), are presented. Among different additive manufacturing systems, inkjet 3DP offers some unique advantages in terms of simplicity of the process, design freedom, used powder recycling, and cost competitiveness. This makes inkjet 3DP a likely method to fabricate parts for applications in the engineering and biomedical sectors.

References

1.	Standard Terminology for Additive Manufacturing Technologies, in ASTM F2792-12a. 2015.

2. Guo, N. and M. Leu, Additive manufacturing: technology, applications and research needs. Frontiers of Mechanical Engineering, 2013. 8(3): p. 215-243. https://doi.org/10.1007/s11465-013-0248-8

3. Nickels, L., AM and aerospace: an ideal combination. Metal Powder Report, 2015. 70(6): p. 300-303. https://doi.org/10.1016/j.mprp.2015.06.005

4. Schiller, G.J. Additive manufacturing for Aerospace. in Aerospace Conference, 2015 IEEE. 2015.

5. Bose, S., S. Vahabzadeh, and A. Bandyopadhyay, Bone tissue engineering using 3D printing. Materials Today, 2013. 16(12): p. 496-504. https://doi.org/10.1016/j.mattod.2013.11.017

6. Ho, C.M.B., S.H. Ng, and Y.-J. Yoon, A review on 3D printed bioimplants. International Journal of Precision Engineering and Manufacturing, 2015. 16(5): p. 1035-1046. https://doi.org/10.1007/s12541-015-0134-x

7. Gibson, I., D.W. Rosen, and B. Stucker, in Additive Manufacturing Technologies: Rapid Prototyping to Direct Digital Manufacturing. 2010, Springer US: Boston, MA. p. 459. https://doi.org/10.1007/978-1-4419-1120-9

8. Qian, M., et al., Additive manufacturing and postprocessing of Ti-6Al-4V for superior mechanical properties. MRS Bulletin, 2016. 41(10): p. 775-783. https://doi.org/10.1557/mrs.2016.215

9. Hull, C.W., Apparatus for production of three-dimensional objects by stereolithography. 1986, Google Patents.

10. Lewandowski, J.J. and M. Seifi, Metal Additive Manufacturing: A Review of Mechanical Properties, in Annual Review of Materials Research. 2016. p. 151-186. https://doi.org/10.1146/annurev-matsci-070115-032024

11. K. Ek, Additive Manufactured Material, in KTH Industrial Engineering and Management. 2014 KTH Royal Institute of Technology

12. Standards, I.I., ISO/ASTM 52900: Additive manufacturing -- General principles -- Terminology. 2015.

13. Olakanmi, E.O., R.F. Cochrane, and K.W. Dalgarno, A review on selective laser sintering/melting (SLS/SLM) of aluminium alloy powders: Processing, microstructure, and properties. Progress in Materials Science, 2015. 74: p. 401-477. https://doi.org/10.1016/j.pmatsci.2015.03.002

14.   Sercombe, T.B. and X. Li, Selective laser melting of aluminium and aluminium metal matrix composites: Review. Materials Technology, 2016. 31(2): p. 77-85. https://doi.org/10.1179/1753555715Y.0000000078

15.   Manakari, V., G. Parande, and M. Gupta, Selective Laser Melting of Magnesium and Magnesium Alloy Powders: A Review. Metals, 2017. 7(1): p. 2. https://doi.org/10.3390/met7010002

16.   Zhang, L.-C. and H. Attar, Selective Laser Melting of Titanium Alloys and Titanium Matrix Composites for Biomedical Applications: A Review Advanced Engineering Materials, 2016. 18(4): p. 463-475. https://doi.org/10.1002/adem.201500419

17.   Zhang, L.C., et al., Review on manufacture by selective laser melting and properties of titanium based materials for biomedical applications. Materials Technology, 2016. 31(2): p. 66-76. https://doi.org/10.1179/1753555715Y.0000000076

18.   Wang, X., X. Gong, and K. Chou, Review on powder-bed laser additive manufacturing of Inconel 718 parts. Proceedings of the Institution of Mechanical Engineers, Part B: Journal of Engineering Manufacture. 0(0): p. 0954405415619883. https://doi.org/10.1179/1753555715Y.0000000076

19.   Sing, S.L., et al., Laser and electron-beam powder-bed additive manufacturing of metallic implants: A review on processes, materials and designs. Journal of Orthopaedic Research, 2016. 34(3): p. 369-385. https://doi.org/10.1002/jor.23075

20.   Körner, C., Additive manufacturing of metallic components by selective electron beam melting — a review. International Materials Reviews, 2016. 61(5): p. 361-377. https://doi.org/10.1080/09506608.2016.1176289

21.   Murr, L.E. and S. Li, Electron-beam additive manufacturing of high-temperature metals. MRS Bulletin, 2016. 41(10): p. 752-757. https://doi.org/10.1557/mrs.2016.210

22.   Gu, D.D., et al., Laser additive manufacturing of metallic components: materials, processes and mechanisms. International Materials Reviews, 2012. 57(3): p. 133-164. https://doi.org/10.1179/1743280411Y.0000000014

23.   Murr, L.E., et al., Metal Fabrication by Additive Manufacturing Using Laser and Electron Beam Melting Technologies. Journal of Materials Science and Technology, 2012. 28(1): p. 1-14. https://doi.org/10.1016/S1005-0302(12)60016-4

24. Tradowsky, U., et al., Selective laser melting of AlSi10Mg: Influence of post-processing on the microstructural and tensile properties development. Materials and Design, 2016. 105: p. 212-222. https://doi.org/10.1016/j.matdes.2016.05.066

25. Wang, M., et al., Fabrication and characterization of selective laser melting printed Ti–6Al–4V alloys subjected to heat treatment for customized implants design. Progress in Natural Science: Materials International, 2016. 26(6): p. 671-677. https://doi.org/10.1016/j.pnsc.2016.12.006

26. Deckard, C.R., Method and apparatus for producing parts by selective sintering. 1989, Google Patents.

27. Ghany, K.A. and S.F. Moustafa, Comparison between the products of four RPM systems for metals. Rapid Prototyping Journal, 2006. 12(2): p. 86-94. https://doi.org/10.1108/13552540610652429

28. Syed, W.U.H., A.J. Pinkerton, and L. Li, A comparative study of wire feeding and powder feeding in direct diode laser deposition for rapid prototyping. Applied Surface Science, 2005. 247(1–4): p. 268-276. https://doi.org/10.1016/j.apsusc.2005.01.138

29. Hammeke, A.W., Laser spray nozzle and method. 1988, Google Patents.

30. Dutta, B., et al., Direct metal deposition. Advanced Materials and Processes, 2009. 167(3): p. 29-31.

31. Mudge, R.P. and N.R. Wald, Laser engineered net shaping advances additive manufacturing and repair. Welding Journal (Miami, Fla), 2007. 86(1): p. 44-48.

32. Zhong, M. and W. Liu, Laser surface cladding: The state of the art and challenges. Proceedings of the Institution of Mechanical Engineers, Part C: Journal of Mechanical Engineering Science, 2010. 224(5): p. 1041-1060. https://doi.org/10.1243/09544062JMES1782

33. Dutta, B., et al., Additive manufacturing by direct metal deposition. Advanced Materials and Processes, 2011. 169(5): p. 33-36.

34. Murr, L.E., et al., Fabrication of Metal and Alloy Components by Additive Manufacturing: Examples of 3D Materials Science. Journal of Materials Research and Technology, 2012. 1(1): p. 42-54. https://doi.org/10.1016/S2238-7854(12)70009-1

35. Murr, L.E., et al., Metal Fabrication by Additive Manufacturing Using Laser and Electron Beam Melting Technologies. Journal of Materials Science & Technology, 2012. 28(1): p. 1-14. https://doi.org/10.1016/S1005-0302(12)60016-4

36.   Sun, Z., et al., Selective laser melting of stainless steel 316L with low porosity and high build rates. Materials & Design, 2016. 104: p. 197-204. https://doi.org/10.1016/j.matdes.2016.05.035

37.   Yadroitsev, I. and I. Yadroitsava, Evaluation of residual stress in stainless steel 316L and Ti6Al4V samples produced by selective laser melting. Virtual and Physical Prototyping, 2015. 10(2): p. 67-76. https://doi.org/10.1080/17452759.2015.1026045

38.   Kok, Y., et al., Fabrication and microstructural characterisation of additive manufactured Ti-6Al-4V parts by electron beam melting. Virtual and Physical Prototyping, 2015. 10(1): p. 13-21. https://doi.org/10.1080/17452759.2015.1008643

39.   Rafi, H.K., et al., Microstructures and Mechanical Properties of Ti6Al4V Parts Fabricated by Selective Laser Melting and Electron Beam Melting. Journal of Materials Engineering and Performance, 2013. 22(12): p. 3872-3883. https://doi.org/10.1007/s11665-013-0658-0

40.   Vastola, G., et al., Modeling the Microstructure Evolution During Additive Manufacturing of Ti6Al4V: A Comparison Between Electron Beam Melting and Selective Laser Melting. JOM, 2016. 68(5): p. 1370-1375. https://doi.org/10.1007/s11837-016-1890-5

41.   Wang, H., et al., A Comparison of Biocompatibility of a Titanium Alloy Fabricated by Electron Beam Melting and Selective Laser Melting. PLOS ONE, 2016. 11(7): p. e0158513. https://doi.org/10.1371/journal.pone.0158513

42.   Zhao, X., et al., Comparison of the microstructures and mechanical properties of Ti–6Al–4V fabricated by selective laser melting and electron beam melting. Materials & Design, 2016. 95: p. 21-31. https://doi.org/10.1016/j.matdes.2015.12.135

43.   Gong, H., et al., Influence of defects on mechanical properties of Ti–6Al–4 V components produced by selective laser melting and electron beam melting. Materials & Design, 2015. 86: p. 545-554. https://doi.org/10.1016/j.matdes.2015.07.147

44.   Toh, W., et al., Microstructure and Wear Properties of Electron Beam Melted Ti-6Al-4V Parts: A Comparison Study against As-Cast Form. Metals, 2016. 6(11): p. 284. https://doi.org/10.3390/met6110284

45.   Karunakaran, K.P., et al., Low cost integration of additive and subtractive processes for hybrid layered manufacturing. Robotics and Computer-Integrated Manufacturing, 2010. 26(5): p. 490-499. https://doi.org/10.1016/j.rcim.2010.03.008

46.   Ye, Z.P., et al., Study of hybrid additive manufacturing based on pulse laser wire depositing and milling. International Journal of Advanced Manufacturing Technology, 2017. 88(5-8): p. 2237-2248. https://doi.org/10.1007/s00170-016-8894-8

47.   Mok, S.H., et al., Deposition of Ti-6Al-4V using a high power diode laser and wire, Part I: Investigation on the process characteristics. Surface and Coatings Technology, 2008. 202(16): p. 3933-3939. https://doi.org/10.1016/j.surfcoat.2008.02.008

48.   Baufeld, B., E. Brandl, and O. van der Biest, Wire based additive layer manufacturing: Comparison of microstructure and mechanical properties of Ti–6Al–4V components fabricated by laser-beam deposition and shaped metal deposition. Journal of Materials Processing Technology, 2011. 211(6): p. 1146-1158. https://doi.org/10.1016/j.jmatprotec.2011.01.018

49.   Liu, Q., et al., Wire feeding based laser additive manufacturing TC17 titanium alloy. Materials Technology, 2016. 31(2): p. 108-114. https://doi.org/10.1179/1753555715Y.0000000075

50.   Geng, H., et al., Optimization of wire feed for GTAW based additive manufacturing. Journal of Materials Processing Technology, 2017. 243: p. 40-47. https://doi.org/10.1016/j.jmatprotec.2016.11.027

51.   Ding, D., et al., Wire-feed additive manufacturing of metal components: technologies, developments and future interests. The International Journal of Advanced Manufacturing Technology, 2015. 81(1): p. 465-481. https://doi.org/10.1007/s00170-015-7077-3

52.   Xiong, J., et al., Fabrication of inclined thin-walled parts in multi-layer single-pass GMAW-based additive manufacturing with flat position deposition. Journal of Materials Processing Technology, 2017. 240: p. 397-403. https://doi.org/10.1016/j.jmatprotec.2016.10.019

53.   Sachs, E.M., et al., Three-dimensional printing techniques. 1993, Google Patents.

54.   Gill, S.S. and M. Kaplas, Efficacy of powder-based three-dimensional printing (3DP) technologies for rapid casting of light alloys. The International Journal of

Advanced Manufacturing Technology, 2011. 52(1): p. 53-64.
https://doi.org/10.1007/s00170-010-2716-1

55. Kirkland, N.T., et al., Synthesis and properties of topologically ordered porous magnesium. Materials Science and Engineering: B, 2011. 176(20): p. 1666-1672. https://doi.org/10.1016/j.mseb.2011.04.006

56. Mun, J., et al., Indirect additive manufacturing based casting of a periodic 3D cellular metal – Flow simulation of molten aluminum alloy. Journal of Manufacturing Processes, 2015. 17: p. 28-40. https://doi.org/10.1016/j.jmapro.2014.11.001

57. De Gans, B.J., P.C. Duineveld, and U.S. Schubert, Inkjet printing of polymers: State of the art and future developments. Advanced Materials, 2004. 16(3): p. 203-213. https://doi.org/10.1002/adma.200300385

58. Derby, B., Additive Manufacture of Ceramics Components by Inkjet Printing. Engineering, 2015. 1(1): p. 113-123. https://doi.org/10.15302/J-ENG-2015014

59. http://www.voxeljet.de/en/systems/vx4000/.

60. Chia, H.N. and B.M. Wu, Recent advances in 3D printing of biomaterials. Journal of Biological Engineering, 2015. 9: p. 4. https://doi.org/10.1186/s13036-015-0001-4

61. Li, Y. and D. Gu, Parametric analysis of thermal behavior during selective laser melting additive manufacturing of aluminum alloy powder. Materials & Design, 2014. 63: p. 856-867. https://doi.org/10.1016/j.matdes.2014.07.006

62. Thompson, S.M., et al., An overview of Direct Laser Deposition for additive manufacturing; Part I: Transport phenomena, modeling and diagnostics. Additive Manufacturing, 2015. 8: p. 36-62. https://doi.org/10.1016/j.addma.2015.07.001

63. Shen, N. and K. Chou. Thermal modeling of electron beam additive manufacturing process: Powder sintering effects. in ASME 2012 International Manufacturing Science and Engineering Conference collocated with the 40th North American Manufacturing Research Conference and in participation with the International Conference on Tribology Materials and Processing. 2012. American Society of Mechanical Engineers.

64. Körner, C., Additive manufacturing of metallic components by selective electron beam melting—a review. International Materials Reviews, 2016: p. 1-17. https://doi.org/10.1080/09506608.2016.1176289

65. Allen, S.M. and E.M. Sachs, Three-dimensional printing of metal parts for tooling and other applications. Metals and Materials, 2000. 6(6): p. 589-594. https://doi.org/10.1007/BF03028104

66. Wiria, F.E., et al., Printing of Titanium implant prototype. Materials & Design, 2010. 31, Supplement 1: p. S101-S105. https://doi.org/10.1016/j.matdes.2009.12.050

67. http://www.exone.com/.

68. Utela, B., et al., A review of process development steps for new material systems in three dimensional printing (3DP). Journal of Manufacturing Processes, 2008. 10(2): p. 96-104. https://doi.org/10.1016/j.jmapro.2009.03.002

69. Kruth, J.-P., et al. Part and material properties in selective laser melting of metals. in Proceedings of the 16th international symposium on electromachining. 2010.

70. Kruth, J.-P., et al., Consolidation phenomena in laser and powder-bed based layered manufacturing. CIRP Annals-Manufacturing Technology, 2007. 56(2): p. 730-759. https://doi.org/10.1016/j.cirp.2007.10.004

71. Olakanmi, E., R. Cochrane, and K. Dalgarno, A review on selective laser sintering/melting (SLS/SLM) of aluminium alloy powders: Processing, microstructure, and properties. Progress in Materials Science, 2015. 74: p. 401-477. https://doi.org/10.1016/j.pmatsci.2015.03.002

72. Yap, C.Y., et al., Review of selective laser melting: Materials and applications. Applied Physics Reviews, 2015. 2(4).

73. Kumar, A., et al., Biocompatibility and mechanical behaviour of three-dimensional scaffolds for biomedical devices: Process-structure-property paradigm. International Materials Reviews, 2016. 61(1): p. 20-45. https://doi.org/10.1080/09506608.2015.1128310

74. Maleksaeedi, S., et al., Toward 3D Printed Bioactive Titanium Scaffolds with Bimodal Pore Size Distribution for Bone Ingrowth. Procedia CIRP, 2013. 5: p. 158-163. https://doi.org/10.1016/j.procir.2013.01.032

75. Butscher, A., et al., Structural and material approaches to bone tissue engineering in powder-based three-dimensional printing. Acta Biomaterialia, 2011. 7(3): p. 907-920. https://doi.org/10.1016/j.actbio.2010.09.039

76. Karageorgiou, V. and D. Kaplan, Porosity of 3D biomaterial scaffolds and osteogenesis. Biomaterials, 2005. 26(27): p. 5474-5491. https://doi.org/10.1016/j.biomaterials.2005.02.002

77.   Artel, A., et al., An agent-based model for the investigation of neovascularization within porous scaffolds. Tissue Eng Part A, 2011. 17(17-18): p. 2133-41. https://doi.org/10.1089/ten.tea.2010.0571

78.   Schaefer, S., et al., How Degradation of Calcium Phosphate Bone Substitute Materials is influenced by Phase Composition and Porosity. Advanced Engineering Materials, 2011. 13(4): p. 342-350. https://doi.org/10.1002/adem.201000267

79.   Choong, Q.L.L.a.C., Three-Dimensional Scaffolds for Tissue Engineering Applications: Role of Porosity and Pore Size Tissue Engineering Part B: Reviews, 2013. 19 (6): p. 485-502. https://doi.org/10.1089/ten.teb.2012.0437

80.   Challis, V.J., et al., High specific strength and stiffness structures produced using selective laser melting. Materials & Design, 2014. 63: p. 783-788. https://doi.org/10.1016/j.matdes.2014.05.064

81.   Liu, Y., et al., Processing and properties of topologically optimised biomedical Ti–24Nb–4Zr–8Sn scaffolds manufactured by selective laser melting. Materials Science and Engineering: A, 2015. 642: p. 268-278. https://doi.org/10.1016/j.msea.2015.06.088

82.   Attar, H., et al., Mechanical behavior of porous commercially pure Ti and Ti–TiB composite materials manufactured by selective laser melting. Materials Science and Engineering: A, 2015. 625: p. 350-356. https://doi.org/10.1016/j.msea.2014.12.036

83.   Liu, Y., et al., Microstructure, defects and mechanical behavior of beta-type titanium porous structures manufactured by electron beam melting and selective laser melting. Acta Materialia, 2016. 113: p. 56-67. https://doi.org/10.1016/j.actamat.2016.04.029

84.   Yan, C., et al., Advanced lightweight 316L stainless steel cellular lattice structures fabricated via selective laser melting. Materials & Design, 2014. 55: p. 533-541. https://doi.org/10.1016/j.matdes.2013.10.027

CHAPTER 2

Methods for Inkjet 3D Printing of Metals

Abstract

The first step in metal inkjet 3DP is to identify a suitable printing method. Accordingly, in this chapter, inkjet 3DP approaches which have been used to fabricate metallic parts are described. In essence, an interaction of binder materials and metal powder enables each layer of the 3D object to create and bond to the previously formed layer. According to the location of the binder materials, printing approaches are classified and advantages and disadvantages of each printing method are presented and discussed.

Keywords

Binder Jetting, Solvent Jetting, Salt Solutions Jetting, Feedstock

2.1 Introduction

Before starting with the fabrication of a part using inkjet 3DP, a printing method must be selected. Accordingly, suitable powder feedstock is prepared and required printing and post processing steps are identified. Fig. 2.1 shows several approaches that have been used for inkjet 3DP. Among these methods, the solvent jetting (SJ) on reactive feedstock method has been only used for ceramic and polymeric materials whereby the feedstock serves as reactants that remarkably react with a solvent, causing interdigitate binding to form a network [1-7]. As result of solvent absorption, the original feedstock may fully convert to other compounds. One of the most intensively studied materials is calcium sulfate hemihydrate, also known as plaster of Paris.

Based on the binder location (see Fig. 2.1), printing approaches are categorized as a binder jetting on powder bed and solvent jetting on powder bed in which binder materials were either selectively deposited in a liquid state by a print head or homogeneously mixed with the metal powder feedstock. The following sections describe methods used in metal inkjet 3DP in detail.

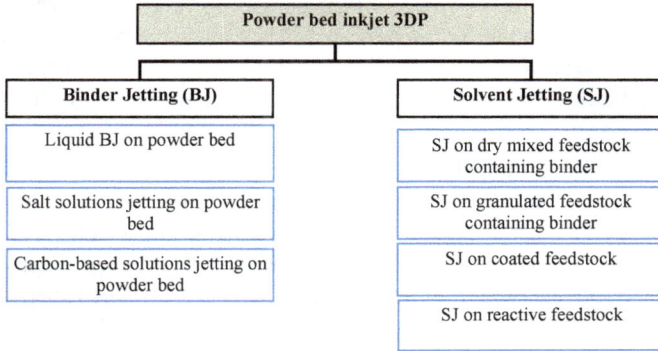

Figure 2.1 Various methods of powder bed inkjet 3DP.

2.2   Binder Jetting (BJ) on Powder Bed

In this approach, metal powder is the only material within the feedstock. Either liquid binders or salt solutions are deposited by a print head in a layer by layer fashion to create the 3D components.

2.2.1   Liquid Binder Jetting on Powder Bed

In this method, the print head deposits a commercially available liquid binder into the powder bed layer-by-layer. In general, no chemical reactions occur between metal powder and the liquid binders within the powder bed. When the print head deposits the binder materials to a layer of powder, an electrical heater may scan the powder bed to partially cure the layer and makes it ready for spreading the subsequent layer. At the end of the build cycle, the as-built part may need to be heated for curing of the binder to enhance the green strength of the part. This is because these binders are mainly made of a thermoset polymer which hardens upon heating. The strength of the as-built parts immediately after removing from the powder bed and before conducting any post-printing process refers to green strength. The curing temperature to fully set the binders varies based on the binder constituents. For example, heating up to 200 °C may be required [8-10]. As the powder bed is gas-permeable, another method for curing the binder is to drive a gas flow through the powder bed which accelerate the curing process [11]. Curing time can be further shortened by passing a gas combined with heating or cooling the powder bed. Fig. 2.2 shows a cross-sectional micrograph of as-built 420 stainless steel part printed using an aqueous binder. Fig. 2.3 displays as-built 3D mesh structures printed by four different sizes of TiNiHf powder after the binder curing at 170 °C for an hour in an oven [9]. As can be seen, small features can be created by the liquid binder jetting on the powder bed method. In particular, using finer metal powder makes it possible to achieve better surface finish and dimensional accuracy in the as-build structures.

The main advantage of the liquid BJ on powder bed method is that binders are compatible with a broad variety of powder materials. Thus, new materials might be developed with a minimal change in printing process parameters [12]. With this view, several models of commercial inkjet 3DP machines have been developed in accordance with this method for fabrication of ceramic, polymeric and metallic materials targeting research or industrial applications [13, 14]. The main drawback of this method is that a debinding step is required to progressively burn out the binder materials before sintering the printed part. Several problems are associated with the debinding step which will be discussed in chapter 4.

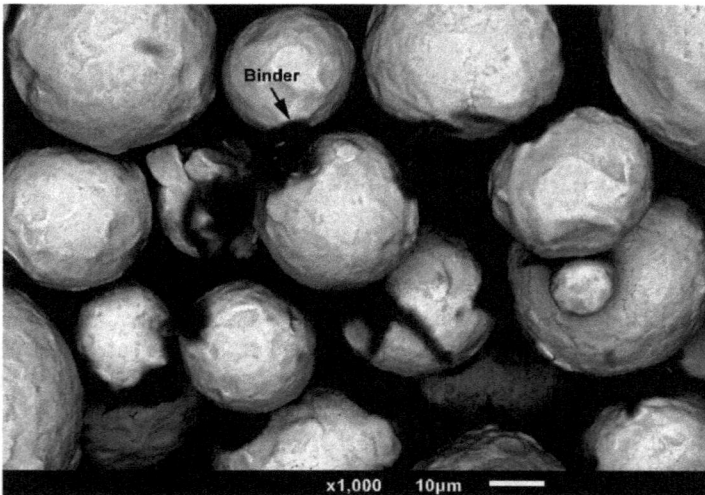

Figure 2.2 SEM image of samples made of 420 stainless steel printed using the liquid BJ on powder bed method.

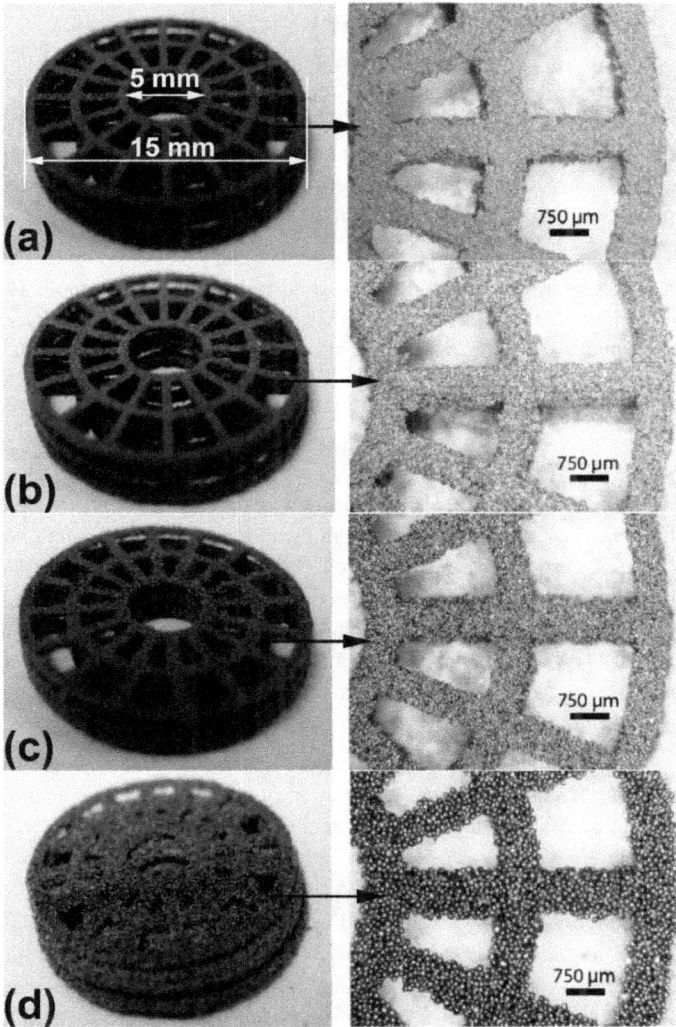

Figure 2.3 As-built mesh structures made of Ti35Ni50Hf15 powder using the liquid BJ on powder bed method with binder saturation level of 170%: a) particle size less than 20 μm, (b) particle size 20–45 μm, (c) particle size 45–75 μm, and (d) particle size 75–150 μm [9].

2.2.2   Salt Solutions Jetting on Powder Bed [15]

Instead of administrating a polymeric binder, a solution of suitable salts, in general metal salt solutions, with desired concentration is deposited by the print head onto the powder bed which only consists of powder with no binder. As a result of an intermediate step of drying and/or heat treatment, a compound originating from the salt solutions creates interparticle necks between adjacent particles and mechanically joins them. Binding mechanisms of different classes of these salts are discussed in detail in the forthcoming sub-sections.

Elimination of the debinding step is the key advantage of this method. However, in addition to the limitations in terms of suitable salts and powder combinations, the liquid reliability is a matter of concern for this method because the liquid with suspended components may clog the print head's nozzles and cause the size of drops or ejection direction of the nozzles to change [16, 17].

2.2.2.1   Recrystallizing Salt

In this class of salts solutions, salt particles are drawn by capillary force to the gap between adjacent powder particles after administration of the salt solutions and upon drying of the liquid carrier. These salts particles bond metal particles to each other within the powder bed. Feedstock spreading along with the salt solution deposition is repeated until the build cycle ends. For example, when 0.7 molar aqueous solution of Rochelle salt ($KNaC_4H_4O_6 \cdot 4H_2O$) is deposited onto a layer of stainless steel within the powder bed, the salt produces a strong body after drying. Interestingly, densification steps such as sintering and infiltration may not be necessary since the recrystallized salt imparts enough strength to the as-built parts.

In addition to limited choice of salts, degradation of printed parts by moisture with time due to re-dissolution of deposited salts is the main drawback of this method.

2.2.2.2   Reducing Salt

In contrast to the recrystallizing salt, printed parts are not removed from the powder bed after the end of the build cycle in inkjet 3DP using the reducing salt solution. At this point, heat treating steps are performed either in a reducing or an inert atmosphere to reduce a metallic constituent of the salts into a metal. The reduced metallic constituent preferably forms a superficial layer on the metal particles within the powder bed and joins them together. Since the heat treatment is conducted below the sintering temperature of the metal particles, the loose particles within the powder bed can be removed allowing the parts' retrieval. It should be noted that the reduced metallic constituents can be sintered or even melted during the heat treatment step. Silver nitrate $(AgNO_3)$, copper nitrate $(Cu(NO_3)_2)$, nickel nitrate $(Ni(NO_3)_2)$, and nickel(II) acetate $(Ni(CH_3COO)_2.4H_2O)$ are some examples of reducing salts.

A perfect support to the printed parts provided by the loose powder in the powder bed during the heat treatment step is the main advantage of this method. Wider applications of this approach are hindered by certain requirements for salts selection which include:

i.   to have high solubility in a liquid carrier such as alcohol or water,

ii.   to be able to reduce to metallic elements and gaseous by-products at temperatures below sintering temperature of the powder particles, and

iii.   the ability to roll out like a superficial film upon reduction to metallic elements.

In addition to the restrictions associated with the salt requirements, some processing challenges might also be confronted in this method. For example, when a reducing atmosphere is required, transporting gas into and out of the printed parts can be a technical concern.

2.2.2.3   Displacing Salt

In this method a salt solution is deposited on the metal powder bed. Following that partial dissolution of the metal powder within the powder bed provides electrons to the salt solution to enable electrochemical displacement reactions between the metallic element in the salt and the powder in the bed. As such, a thin film of the metal originated from the salts is deposited all over the metal powder, making particles stick together by formation of interparticle bridges once the solution dries. During subsequent heat treatment, these bridges could reduce to metallic constituents containing metallic elements of both the salts and the powder bed. These constituents show structures of dispersed particles, alloys, composite etc. For example, silver carbonate (Ag_2CO_3) solution with the concentration of 0.3 molar in the solvent of 87 vol.% water and 13 vol.% ammonium hydroxide(NH_3) was utilized for a powder bed made with molybdenum (Mo). Electrons transferred from Mo to the solution, and the metallic cations of the salt converted to the metallic element (i.e. Ag) upon deposition of the solvent on the powder bed. After heat treatment at 980 °C, silver films formed and covered the Mo particles. Then, the part was removed from the powder bed.

Enabling an electrochemical reaction between salt solutions and the powder bed is the key consideration for selection of suitable displacing salts. In this respect, the electromotive force series can be taken into account as a useful guideline to narrow down possible combinations.

2.3 Solvent Jetting (SJ) on Powder Bed

An alternative approach to the BJ approach is to use an incorporated dry binder component in the feedstock which binds the metal powder upon interacting with a selectively deposited liquid solvent by the print head. Higher deposition reliability and the ratio of binder to metal powder in the as-built parts can be considered as advantages of this approach since there are fewer rheological constraints in solvent jetting compared to the binder jetting [18]. However, issues linked to the debinding step still exist. Additionally, pre-processing steps must be carried out to translate metal powder into feedstock possessing acceptable properties. Characteristics of feedstock materials will be discussed in chapter 3. Different techniques that have been used to prepare the feedstock for the SJ on powder bed approach are:

i.   dry mixing of binder and metal powder,

ii. coating of metal powder with binder, and

iii. making granules from binder and metal powder.

2.3.1 Solvent Jetting on Dry Mixed Feedstock

In this method, solid binder powder such as polyvinyl alcohol (PVA) is dry mixed with metal powder using a vibratory mill or ball mill mixer in the presence of balls to facilitate the mixing process and to form a uniformly dispersed feedstock mixture. The powder mixture is then sieved to filter out the agglomerated particles. This mixture is then fed to an inkjet 3DP machine. Fig. 2.4 shows a representative of such feedstock. When a water-based liquid solvent, for example, deionized water together with methanol in small quantities, is selectively deposited by the print head, the solvent partially or fully dissolves the binder particles and causes the binding process to initiate. As such, these binder particles stick to neighboring metal particles to fabricate parts as shown in Fig. 2.5.

As can be seen from Fig. 2.4, a typical powder feedstock comprises of metal particles with spherical shape and binder material having irregular shape. It has been demonstrated that mixtures consisting of different particle shapes are more readily segregated compared with mixtures with similar particle shapes [19, 20]. Furthermore, segregation is remarkably higher for a binary mixture which composes of spherical and irregular shaped particles [21]. Moreover, in the experience of the authors, this segregation is a serious issue when it comes to the fabrication of large parts or series production. Issues related to segregation of powder are discussed in detail in chapter 3.

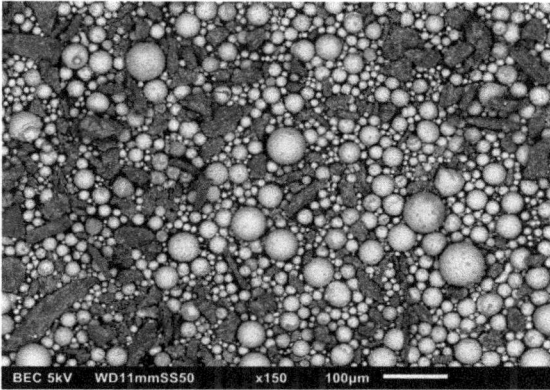

Figure 2.4 A dry mixed Titanium (Ti) powder (spherical particles) with 20 vol. % PVA (irregular particles) ball milled for an hour.

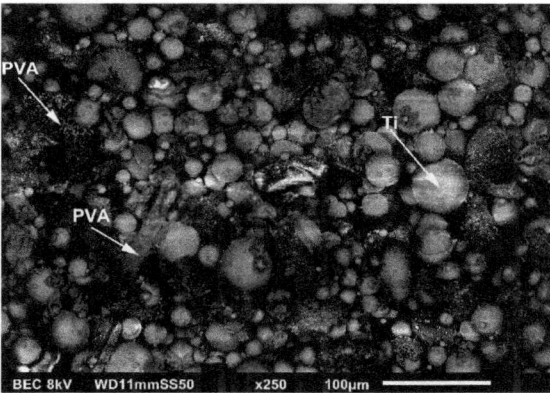

Figure 2.5 Cross-sectional view of as-built parts made of Ti with 20 vol. % PVA as a binder.

2.3.2 Solvent Jetting on Coated Feedstock

In this method, metal powder is coated with a water-soluble polymer and then a liquid solvent is selectively deposited by the print head to stick the powder together. For example, 1 wt.% of water soluble starch can be mixed with 99 wt.% of metal powder and ball milled to prevent the powder from agglomeration during the coating treatment [22].

Fig. 2.6 presents the surface appearances of the coated metal powder with starch. As can be seen, starch particles partially adhered to the metal powder for a short coating time of one hour, while the total coverage of the entire surface was realized after three hours of milling. Increasing the coating time caused agglomerated particles to become greater than 20-fold of the initial size because starch easily cohered with each other [22]. Screening should be used to de-agglomerate these particles.

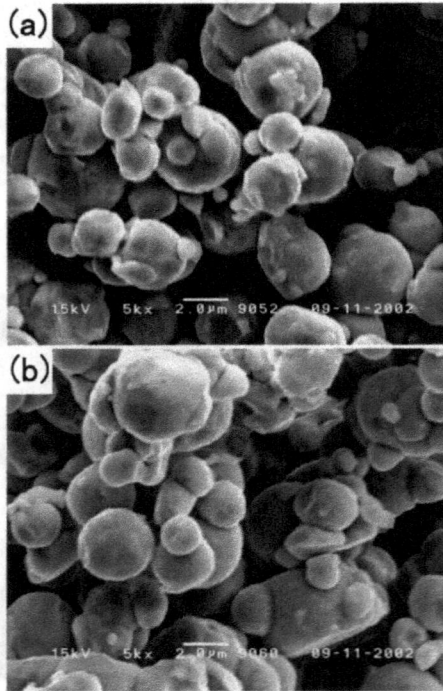

Figure 2.6 The appearance of coated nickel powders with starch: a) feedstock after ball-milling for an hour and b) feedstock after ball-milling for three hours [22].

2.3.3 Solvent Jetting on Granulated Feedstock

Similar to the aforementioned SJ techniques, this method is based on the deposition of solvent droplets on the granules of metal powder and binder materials which locally binds the granule to form a layer as schematically depicted in Fig. 2.7. A typical granule consists of metal powder together with a thermoplastic binder. They can be prepared by wet blending, drying, milling, and sieving [23]. More specifically, irregular or spherical shaped metal particles with size between 0.1 and 100 μm can be granulated with a polymeric binder which is soluble in alcohol, water or ketones, such as PVA, polyvinyl butyral (PVB), and polyvinyl acetate (PVAC). This method generates less dust, and granules flow more smoothly while being spread by the roller.

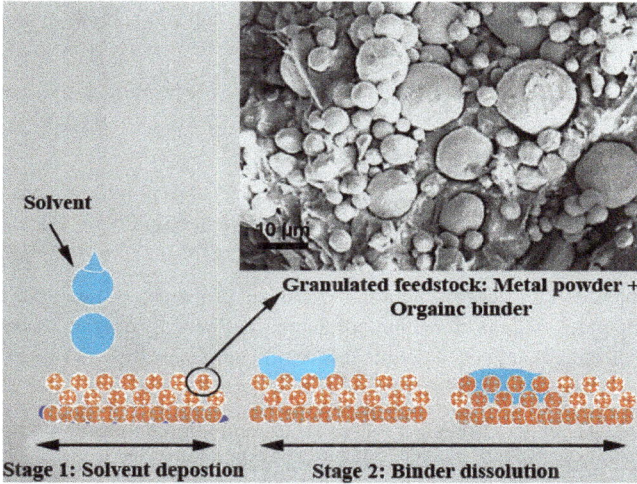

Figure 2.7 The principle of the SJ on a granulated feedstock method.

Summary

This chapter summarizes the printing methods which have been utilized up to date for inkjet 3DP of metals. No preprocessing is required for the binder jetting on powder bed approach as metal powder is the only material in the powder bed. However, feedstock preparation steps are required for the solvent jetting on powder bed approach since binder materials must be mixed with the metal powder. With the exception of the salt solutions jetting on powder bed method, all printing methods make use of polymeric materials to bind metal powder to each other. Accordingly, a debinding step is required. Among the different methods, the liquid binder jetting on powder bed method is widely used in industrial scale.

References

1. Butscher, A., et al., Structural and material approaches to bone tissue engineering in powder-based three-dimensional printing. Acta Biomaterialia, 2011. 7(3): p. 907-920. https://doi.org/10.1016/j.actbio.2010.09.039

2. Wang, T., R. Patel, and B. Derby, Manufacture of 3-dimensional objects by reactive inkjet printing. Soft Matter, 2008. 4(12): p. 2513-2518. https://doi.org/10.1039/b807758d

3. Lam, C.X.F., et al., Scaffold development using 3D printing with a starch-based polymer. Materials Science and Engineering: C, 2002. 20(1): p. 49-56. https://doi.org/10.1016/S0928-4931(02)00012-7

4. Giordano, R.A., et al., Mechanical properties of dense polylactic acid structures fabricated by three dimensional printing. J Biomater Sci Polym Ed, 1996. 8(1): p. 63-75. https://doi.org/10.1163/156856297X00588

5. Suwanprateeb, J., Improvement in mechanical properties of three-dimensional printing parts made from natural polymers reinforced by acrylate resin for biomedical applications: a double infiltration approach. Polymer International, 2006. 55(1): p. 57-62. https://doi.org/10.1002/pi.1918

6. Gbureck, U., et al., Direct Printing of Bioceramic Implants with Spatially Localized Angiogenic Factors. Advanced Materials, 2007. 19(6): p. 795-800. https://doi.org/10.1002/adma.200601370

7. Khalyfa, A., et al., Development of a new calcium phosphate powder-binder system for the 3D printing of patient specific implants. Journal of Materials Science: Materials in Medicine, 2007. 18(5): p. 909-916. https://doi.org/10.1007/s10856-006-0073-2

8. Bai, Y. and C.B. Williams, An exploration of binder jetting of copper. Rapid Prototyping Journal, 2015. 21(2): p. 177-185. https://doi.org/10.1108/RPJ-12-2014-0180

9. Lu, K., M. Hiser, and W. Wu, Effect of particle size on three dimensional printed mesh structures. Powder Technology, 2009. 192(2): p. 178-183. https://doi.org/10.1016/j.powtec.2008.12.011

10. Ziaee, M., E.M. Tridas, and N.B. Crane, Binder-Jet Printing of Fine Stainless Steel Powder with Varied Final Density. JOM, 2017. 69(3): p. 592-596. https://doi.org/10.1007/s11837-016-2177-6

11. MCCOY, M.J. and T. Lizzi, Methods and apparatuses for curing three-dimensional printed articles. 2016, Google Patents.

12. Utela, B., et al., A review of process development steps for new material systems in three dimensional printing (3DP). Journal of Manufacturing Processes, 2008. 10(2): p. 96-104. https://doi.org/10.1016/j.jmapro.2009.03.002

13. http://www.exone.com/

14. http://www.voxeljet.com/

15. Sachs, E.M., et al., Metal and ceramic containing parts produced from powder using binders derived from salt. 2003.

16. Calvert, P., Inkjet Printing for Materials and Devices. Chemistry of Materials, 2001. 13(10): p. 3299-3305. https://doi.org/10.1021/cm0101632

17. Calvert, P. and T. Boland, Biopolymers and Cells, in Inkjet Technology for Digital Fabrication. 2012, John Wiley & Sons, Ltd. p. 275-305. https://doi.org/10.1002/9781118452943.ch12

18. Bredt, J.F. and T. Anderson, Method of three dimensional printing. 1999, Google Patents.

19. Swaminathan, V. and D.O. Kildsig, Polydisperse powder mixtures: Effect of particle size and shape on mixture stability. Drug Development and Industrial Pharmacy, 2002. 28(1): p. 41-48. https://doi.org/10.1081/DDC-120001484

20. Johanson, J.R., Predicting Segregation of Bimodal Particle Mixtures Using the Flow Properties of Bulk Solids. Pharmaceutical Technology, 1996. 20(5): p. 46-57.

21. Tang, P. and V.M. Puri, Methods for Minimizing Segregation: A Review. Particulate Science and Technology, 2004. 22(4): p. 321-337. https://doi.org/10.1080/02726350490501420

22. Kakisawa, H., et al., Dense P/M Component Produced by Solid Freeform Fabrication (SFF). Materials Transactions, JIM, 2005. 46(12): p. 2574-2581. https://doi.org/10.2320/matertrans.46.2574

23. Carreno-Morelli, E., S. Martinerie, and J.E. Bidaux, Agglomerate for 3D-binder printing. 2007.

CHAPTER 3

Material Selection Guidelines for Inkjet 3D Printing

Abstract

Currently no description is available for materials selection criteria for the fabrication of metallic or ceramic components using inkjet 3DP. This chapter attempts to introduce key material selection criteria that should be taken into account for a successful build cycle of a part including its post-printing steps. Four factors, namely, powder flowability, powder packing density, powder segregation and powder sinterability are major requirements for feedstock selection and an overview of some assessment procedures for evaluation of these factors are provided. Furthermore, the influence of the feedstock parameters such as particle size, particle size distribution and particle shape is highlighted. Finally, a general outline of ink selection guidelines is presented.

Keywords

Flowability, Packing Density, Segregation, Sinterability, Binder, Solvent, Inkjet 3DP, Metal, Ceramic

3.1 Introduction

The next step after determining a suitable printing method for inkjet 3DP of material is to select proper materials in terms of:

i. metal powder,

ii. binder, and

iii. solvent

These are selected to ensure that their combinations provide suitable properties for building a part as well as its post-printing steps. In general, powder can be deposited either in a wet or dry state to ascertain that the powder flows easily to form a smooth and uniform thickness layer within a minimum layer formation time. Mechanical spreading of fine powder (< 1 μm) may be difficult as the flow of particles is dominated by interparticle forces. Hence, a liquid carrier is required to ensure their free motion [1, 2]. In metal inkjet 3DP, particle sizes greater than 5 μm, particularly greater than 20 μm, are preferably deposited in the dry state by means of mechanical spreading due to an ease of spreading [3, 4]. Moreover, spherical particle shapes are usually preferred over faceted or irregular powder since they have lower internal friction. They also tend to flow better and provide better wettability with binders [4, 5]. Many other processing variables are influenced by the powder characteristics. Feedstock dependent variables include:

a) Powder and binder interactions.

b) Powder packing density.

c) Printing layer thickness.

d) Layer surface finish.

e) Printing resolution, that is, the smallest or minimum printable feature size.

f) Sinterability.

g) Final density.

The forthcoming sections provide insights into the criteria that should be considered to have a judicious selection of powder feedstock.

3.2 Characteristics of Feedstock Materials

To fabricate reproducible components with predictable properties that attract industry confidence, it is vital to determine the intrinsic properties (e.g. chemical composition, sinterability) and extrinsic properties (e.g. packing density, flowability) of the feedstock powder. The most relevant standard guides specified for AM processes are ASTM F3049 [6] and ISO 17296-3 [7]. They refer to a wide variety of other ASTM, ISO and Metal Powder Industries Federation (MPIF) standard testing procedures that may be useful for all powder-based AM methods encompassing powder bed methods and directed energy deposition. These standard test methods are applicable both for the virgin and used metal powder. These two standards introduce techniques for taking samples, determining powder size, characterizing powder morphology, determining chemical composition, and measuring powder flowability and density. However, test methods for the determination of feedstock's flow characteristic or powder packing density, especially under dynamic conditions, are not entirely addressed given their crucial impact on the build cycle as well as parts' final properties [8, 9]. More importantly, only technical issues associated with conducting standard procedures are described in the standard guides, not the influential parameters or mechanisms. These four factors (powder flowability, packing density, sinterability and segregation) are thoroughly discussed under this section.

3.2.1. Powder Flowability

Flowability describes the capability of powder to flow under loading and rearrange itself in static states. Sufficient powder flowability enables the roller to spread thin and homogenous layers of powder, which enhance printing resolution, dimensional accuracy, and density of a final part. Poor flowability leads to difficulties in feedstock spreading and non-uniformity in powder bed distribution, in addition to uneven powder bed's surface. Thus, flowability of powder was highlighted as the first consideration for developing new materials for powder bed AM systems [10].

According to Prescott and Barnum [11], powder flowability is a multi-faceted issue and properties of materials, storing, handling, and processing of materials are influential factors affecting powder flowability. In order to determine powder flow behavior, a couple of characterization methods are available. Surprisingly, there might be inconsistency in results of these different flowability techniques when it comes to comparison of one to another [12, 13]. Due to close connections between stress states and flowability, flow characteristics measured under different stress states are not interchangeable [14, 15]. In other words, it is not recommended to extend results of a test method subjected to a static stress state to a dynamic situation and vice versa [15]. Thus, the key consideration is that the powder characterization methods should be as similar as possible to the real processing that the feedstock powder will undergo [14, 15].

To the authors' knowledge, there is no established quantitative method in the field of powder bed AM techniques to determine or predict feedstock flow behavior under the condition of feedstock spreading by a roller. The current approach of powder development is mainly based on trial and error, but quantitative approaches are required to reduce feedstock's development time cycle. Several test methods which require either a readily accessible apparatus or highly sophisticated ones are used to quantify flow characteristics under static and dynamic conditions [16]. Angle of Repose (AoR) test, which is a well-established and simple method, consists of both dynamic and static elements in which the pile of powder is formed onto a flat surface by means of pouring powder through a funnel as depicted in Fig. 3.1a [17]. AoR is the angle between the horizontal axis and a tangent to the surface of a cone-shaped pile, which arise from the equilibrium condition between kinetic energies of the powder falling from the funnel and static interparticle forces between particles forming an stationary pile. A lower AoR signifies lower interparticle forces, indicating higher powder flowability.

In addition, the ratio of tap to bulk density, called Hausner ratio (HR), provides a statement on powder flowability. Compressibility index, also known as Carr's index, is another parameter which can be derived from the tap and bulk density differences by following the equation [16].

$$K_I = (\text{tapped density} - \text{bulk density}) / \text{tapped density} \qquad (3\text{-}1)$$

Either Carr's index of zero or HR value of one indicates an incompressible powder, interpreting the best possible powder flowability. HR value should be smaller than 1.25 for powder to be labeled as a suitable feedstock for powder bed AM systems [18].

AoR, HR or Carr's index tests can be served as simple indicators for acquiring first-hand information on flowability. However, the stress states under these tests differ significantly from the actual process condition in which the feedstock is spread along the powder bed by means of the roller. Furthermore, reproducibility is the major concern for these powder characterization methods. Dynamic powder testing, which involves evaluating flow properties while test materials are in motion, is a new characterization method to evaluate a newly developed feedstock under an analogous stress state to the real practice in which the roller spreads feedstock over the powder bed. Revolution Powder Analyzer is a commercially available instrument for dynamic analysis of powder that includes a rotating drum with two transparent sides, a digital camera, an image acquisition system, and software as depicted in Fig. 3.1b [19]. The behavior of powder inside the rotating drum is captured by the camera and analyzed with the aid of the software. Up to 20 parameters associated with the behavior of powder inside the drum can be measured. These parameters include avalanche angle, avalanche energy, surface fractal, sample density, volume expansion ratio, fluidized volume slope, fluidized height slope, and final settling time [19, 20]. These parameters can be interpreted and used as the characterization indices of powder and also be correlated to other test methods for powder characterizations [15, 20, 21]. In addition to the close similarity in stress state conditions inside the rotating drum with the actual spreading state in AM machines, rotational speeds of the drum can be adjusted precisely to translational speeds of the roller to mimic the real powder flow characteristics [20]. It should be noted that dynamic analysis complements static analysis, resulting in a more comprehensive understanding of the powder flow behavior.

Another method to analyze the powder flowability in a dynamic situation is to utilize in-house built set-up [22]. Schematic illustration of such a lab-scale device is shown in Fig. 3.1c. As can be seen, the set-up includes a powder spreader, the measurement plate which is a rectangular area with specific dimensions to have an accurate estimation of the powder volume, and a balance to measure the weight of spread powder inside the measurement plate. Accordingly, the ratio between the layer density and the tap density of the powder inside the measurement plate can provide an index for the powder flowability [22]. Adjusting such a set-up allows the use of different types of spreaders (i.e. a counter-rotating roller or a recoater blade) and powder spreading velocities. Moreover, it is possible to assess the powder bed roughness as well as its packing density [22].

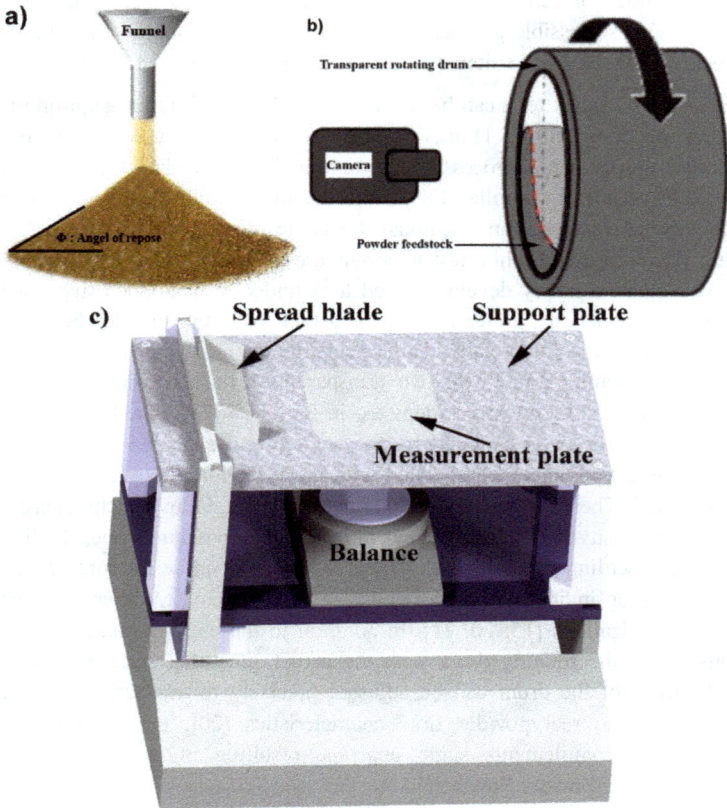

Figure 3.1 Schematic representations of methods for evaluating flow behavior of powder feedstock, a) Angle of Repose test, b) Revolution Powder Analyzer (adapted from [23]) , and c) The mock powder spreader (adapted from [22]).

Powder flowability is closely linked with interparticle forces [24]. Thus, flowability is dependent upon many parameters, namely:

a. Particle size.

b. Size distribution of particles.

c. Shape of particles.

d. Surface morphology of particles.

e. Environmental conditions.

f. Temperature.

g. Chemical composition of powder [8, 25, 26].

The theoretical correlations between different interparticle forces acting on single-point contact and equally sized spherical particles in air are plotted as function of particle size in Fig. 3.2. Irrespective of the characterization techniques (i.e. dynamic or static analysis), flowability reduces remarkably when particle size decreases [15]. As a result of the larger surface areas of fine particles (< 20 µm), there is a higher extent of cohesive and adhesive interparticle forces, which attribute to the capillary, electrostatic, and van der Waals forces [27-29]. When these interparticle forces outweigh the gravitational force and become dominating forces, powder exhibits poor flow behavior and exhibit a high tendency for agglomeration. On the other hand, higher gravitational force together with diminished interparticle forces allow larger particles to readily roll over each other and thus flow more freely. Despite the fact that the prediction of flowability based on a particle size distribution is challenging, flowability decreases with increasing width of particle size distribution for powder with the same median particle sizes [30]. It is also obvious that with the similar shape of particle size distributions, flowability diminishes with decreasing median particle size since the number of particles' contacts per unit of area is inversely proportional to the square of the particle size. Particles with rough surface morphology or irregular shape lead to poor flowability. Angular and edgy particles tend to mechanically interlock with each other, causing higher interparticle adhesion and friction, thereby resisting free-flow [31]. Similarly, it is more likely that particles with rough surface hook and twist together than particles with smoother surface morphology [32]. Therefore, particles with high sphericity and smooth surface are favorable. It is not possible to make a general comment regarding the influence of environmental conditions on flowability. Consider the case of moisturized powder, on the one hand, creating moisture bridges between particles may adversely affect flowability, but on the other hand, moisture may act as lubricant to reduce cohesion and can improve the powder flow properties [33].

The only way to enhance powder flowability is to weaken the interparticle forces [24]. One way to counter poor flowability of fine particles is to add some additives to the powder. These additives reduce van der Waals forces by distancing particles from one another [34]. Several other techniques such as adding nano-sized additives, chemical surface treatment of particles, and mixing lubricants can also be employed to diminish interparticle forces [35, 36].

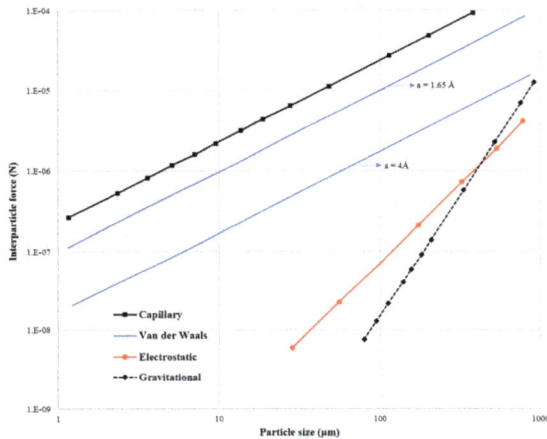

Figure 3.2 The magnitude of various interparticle forces as function of particle size for single-point contact between equally sized sphere in air (adapted from [37]).

3.2.2 Powder Packing Density

Powder packing density (PPD) is the relative density of each layer of the powder bed being spread via the roller. Several investigations indicated that PPD correlates strongly with mechanical, structural and superficial properties of fabricated parts [8, 22, 38]. In particular, PPD is the most influential parameter on mechanical properties, especially the ultimate tensile strength (UTS) and elongation; the closer the particle packing, the higher the UTS and the ductility [8]. This is because higher PPD brings about lesser interparticle voids inside the powder bed and thus reduce the parts' porosity. The direct relationship between mechanical properties and porosity percentages of inkjet 3D printed parts is demonstrated by Spath et al. [39]. In addition to the formation of a rough surface, low PPD leads to unfavorable interaction between feedstock and ink (i.e. binder or solvent) and low green strength of the as-built parts [40]. It is anticipated that powder with suitable flowability shows an efficient packing since PPD and powder flowability are

40

interrelated [24, 41]. The greater the flowability the lower is the particle interaction. Accordingly, almost all aforementioned parameters affecting powder flowability influence PPD as well. Similar to powder flowability, powder with smaller particle size shows lower packing density [26, 40]. However, it is stated that mixture of fine and large particles can create a denser powder bed if the fraction of large particles is higher than that of fine particles within the particle size distribution due to the gap filling effect of the fine particles [8, 26, 42]. In view of this, a broader width of particle size distribution leads to higher PPD because finer particles fill up the pores. Particles with a higher degree of sphericity tend to pack in tighter arrangements which stem from their higher mass density and packing efficiency [43, 44]. Increase in the sphericity of particles causes the void volume to decline when compared to irregular and edgy shaped particles that lead to interparticle bridges due to high friction forces [45, 46]. For example, Fig. 3.3 displays micro-computed tomography (μCT) images of two as-built Inconel 625 parts made of two distinct particle shapes, namely, spherical and irregular shape [47]. These samples were printed using the liquid binder jetting on powder bed method and the average particle size for spherical and irregular shape particles were 32 and 34.5 μm, respectively. According to μCT results, the relative densities of the as-built parts were approximately 50 % and 60% for the irregular and the spherical powder, respectively. 10% higher relative density of spherical powder can be solely attributed to the effect of particle morphology since all the other parameters, including average particle size, particle size distribution and printing parameters, are almost the same.

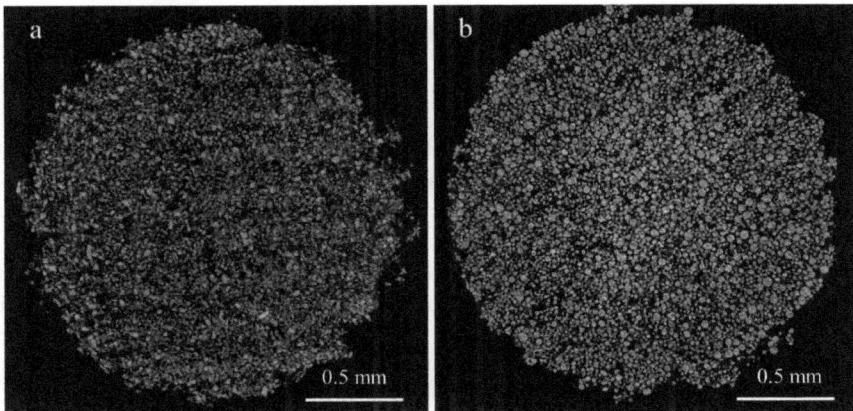

Figure 3.3 μCT images of the as-built Inconel 625 samples: (a) powder with irregular shape and (b) powder with spherical shape (courtesy – A. Mostafaei [47]).

PPD is not only connected with the physical properties of feedstock, it is also related to the parameters of the build cycle. Simulation results of spreading feedstock via the counter-rotating roller systems demonstrated that the coefficient of contact friction between feedstock and the roller affects the PPD. The lower the friction coefficient, the higher the PPD will be [48]. It should be noted that the friction between the roller and the feedstock is dependent upon both the feedstock's physical properties, e.g. particle size, particle shape or powder surface morphology, and the roller related parameters such as its surface roughness or its rotational speed for the feedstock deposition [48]. The roller compacts the powder feedstock to some extent while spreading it over the powder bed, and the narrowest gap between the roller and the underlying feedstock is the location where the maximum compaction pressure occurs [48, 49]. Therefore, the powder layer thickness (discussed in chapter 4) is another effective parameter of PPD. Higher layer thickness causes the PPD to decline [48]. This is because the magnitude of compaction pressure diminishes for a larger layer thickness of the feedstock.

One method to directly quantify the PPD is to use mock devices mimicking the actual spreading situations in the 3DP machine, for example, the lab-scale set-up shown in Fig. 3.1c [35]. The other methods involve the indirect evaluation of PPD. Measuring the bulk density (i.e. the density of powder under nonstandard conditions) together with the tap density, that is, the density of powder in a container tapped according to a specific method [50], provides a valuable quantitative estimation of the PPD. Given that the roller does not notably compress the feedstock within the powder bed, the bulk and tap densities of the feedstock can be considered as the lowest and the highest attainable values for PPD, respectively [22, 40]. More specifically, the packing condition of the powder bed is much closer to a free-flowing situation (i.e. the bulk density) rather than the tapped state because the feedstock mainly compacts under the effect of gravitational force. Hence, it would be anticipated that the PPD is higher than the bulk density of the feedstock and lower than that of the tapped condition, particularly for the case of feedstock with high flowability. Nonetheless, some researchers reported that the PPD agrees with values of the bulk density of powder, whereas others indicated that the green density of the inkjet 3D printed parts matches with value of the tap density [40, 51]. This inconsistency arises from differences in the feedstock flowability and parameters associated with the build cycle.

The interparticle voids will not markedly diminish over the course of feedstock spreading. Thus, selection of feedstock with proper physical properties is of prime importance due to the fact that an increase in the bulk density of the initial powder enhances the PPD. Butscher et al. [36] demonstrated that the bulk density of calcium phosphate powder rose in excess of 65% as a result of fine-tuning the feedstock's physical properties, including particle size, particle size distribution, and morphology. In another example, the bulk density of hydroxyapatite granules was raised to over 20 % by adjusting the particle size [26].

3.2.3 Powder Sinterability

Acceptable powder flowability and good powder packing density are fundamental requirements that must be taken into account during the feedstock powder selection. Even powder possessing sufficient packing and spreading properties might be considered as inappropriate feedstock at times due to other undesirable intrinsic material properties. Given the importance of the sintering process on physical and mechanical properties of the inkjet 3D printed components, sinterability of powder should be considered as a paramount consideration for selecting feedstock materials. Particle size, particle size distribution, particle shape along with powder chemical composition, surface chemistry, and oxide film's thickness determine the powder sinterability [52, 53].

Several studies have been demonstrated that the apparent activation energy for densification that is the minimum energy required for the sinter bonding to occur is remarkably influenced by the particle's size [54, 55]. The activation energy decreases along with the particle's size. For instance, the activation energy for fine tungsten (W) powder (i.e. 0.4–3.0 μm) is 90 kJ/mol , as opposed to 150 kJ/mol for coarse W powder (i.e. 5.0 to 7.0 μm) [55]. This trend arises from higher interaction and contacts area between a pair of particles with decreasing particle size for a given volume, enhancing the diffusion process among particles during sintering. Furthermore, there is a correlation between the particle size and the sintering stress. The sintering stress is always higher for finer particle [55]. Thus, powder sinterability is promoted as particle size reduces and finer powder sinters at a much lower temperature than that of a coarser one. To achieve density of 69% in W samples, for example, the sample made of powder with the particle size of 1 μm was required to be sintered at 1500 °C. In contrast, samples made of 3.0 μm W powder needed to be sintered at 1700 °C to achieve 69% in density [55]. Fig. 3.4 represents impact of sintering temperature and particle size on powder sinterability of inkjet 3D printed Inconel 718 samples. Finer particle size demonstrates greater frequency of neck formation between particles and thus higher densification rate, irrespective of temperatures [56].

It should be pointed out that decreasing melting temperature as a result of decreasing particle size is not valid for micron-sized powder. It is well established theoretically and experimentally that surface-induced melting phenomenon is only applicable when the ratio between the number of atoms in the bulk of particle and its surface is in the thousand range [57-59]. For a particle larger than a critical radius, such a size-dependence phenomenon of melting temperature becomes negligible, and the particle behaves like a solid material [60]. TEM examinations of bismuth, indium, lead, and tin demonstrated that nano-sized particles with a size in the range from 3 to 50 nm tend to melt below melting temperature of their corresponding bulk materials, and their melting temperatures further diminish with a decrease in particles' radius [59].

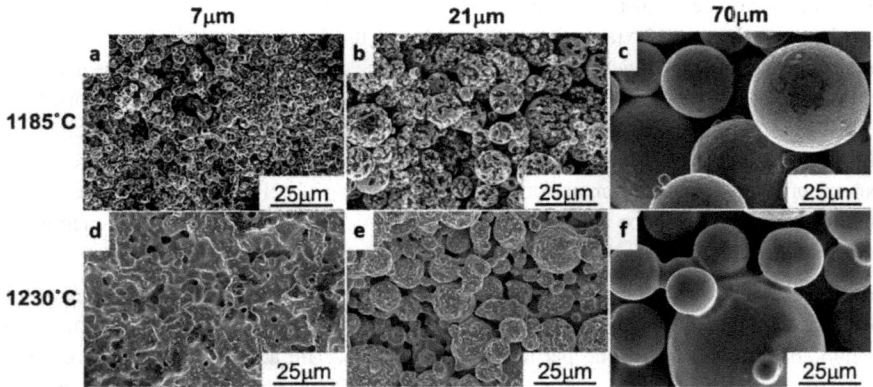

Figure 3.4 influence of sintering temperature and particle size on densification and sinterability of inkjet 3D printed Inconel 718 using the liquid BJ on powder bed method [56].

Unlike the effect of particle size distribution (PSD) on PPD, most experimental, theoretical and analytical studies concur that a narrower PSD is preferred in order to achieve a dense, uniform, and a fine-grained part after sintering [52, 61-64]. Ting and Lin [61, 62] divided the sintering process into two stages; before and after the incidence of grain growth. During the first stage, which is synonymous with the initial stage and a part of the intermediate stage, according to Coble's classification which is discussed in chapter 5, reducing the system energy by mass diffusion from particles to pores is the driving force for sintering. After the occurrence of grain growth, a part of the intermediate stage and the whole of final stage based on Coble, the reduction of grain-boundary energy through grain growth along with the surface energy reduction are the driving forces. Prior to the incidence of grain growth, a wider PSD shows higher sinterability [52, 63]. As mentioned earlier, a broader PSD brings about a higher packing density and therefore the higher average coordination number of particles (nearest-neighbor contacts per particle). In addition, finer particles possess higher sinterability.

This synergy causes finer particles to be consumed rapidly by larger neighboring particles, resulting in enhanced sinterability at the first stage of sintering for a given sample with broad PSD [65, 66]. On the other hand, the depletion of samples from fine particles diminishes both the number and the neck area between the particles, lowering the sinterability of a broad PSD than that of a narrow PSD at the start of the second stage of sintering. This decline in sinterability is attributed to the fact that the neck regions serve as a site for vacancies to be annihilated. Upon further elimination of the neck area during sintering, grain growth by surface diffusion become a dominant mechanism of microstructural evolution, leading to the poorer densification rate for a wider PSD [63]. Despite these two opposing effects of PSD's width on sintering, the overall sinterability of a narrower PSD is always better than that of a broader one [52, 61, 62, 64, 67]. Moreover, the width of PSD was found to be inversely proportional to the densification rate [63].

The better sinterability of a narrower PSD raises the question of how narrow the PSD is required. There is no agreement in the literature on this. Some results are in direct conflict with each other since the effect of PSD on sinterability are influenced with many other factors such as material types, initial density, consolidation methods, and average particle size. Some experimental and simulation works have suggested that monosized particles (i.e. synonymous with narrowest possible PSD) are preferred, showing the highest sinterability and final relative density [63, 68, 69], while some reported better sinterability of compacts having a broader PSD [70, 71]. Ma and Lim [71] carried out an experimental research on the effect of PSD on sinterability of non-agglomerated alumina powder. Their results indicated that there is an optimum range of PSD to obtain best sinterability. The results also showed that all samples with different PSD resulted in better sinterability in comparison with samples of near-monosized PSD. In particular, the sample possessing an optimum PSD sintered to full density at temperature 200 °C lower than that of near-monosized PSD. Despite the existence of the optimum PSD, Ma and Lim [71] eventually recommended using a narrow PSD over the monosized PSD to control the microstructural features and to achieve higher sinterability of the compacts. This is because such an optimum PSD may vary substantially based on the consolidation methods and conditions used, even for a given powder system.

Even though spherical shaped particles provide excellent flowability and PPD, particles with spherical shape may adversely affect sinterability. Taruta et al. [72] demonstrated that the sinterability of spherical alumina particle in the range from 6 to 7 μm was almost the same as compared with the plate-like powder with similar size. On the other hand, it demonstrated that irregularly shaped glass or aluminum particles offer better sinterability than spherical ones [53, 73]. In case of glass, better sinterability was attributed to much smaller effective particle size of irregular shape which could be one-fifth of similar spherical particles. And for Al particles, the ease of breaking surface oxide layers during sintering was the root cause of better sinterability.

It should be pointed out that the aforementioned trends are generic for the case of conventional sintering processes which are currently dominated in AM technologies. A slight change or reversed trends might be observed in the powder sinterability using unconventional sintering techniques such as microwave sintering and spark plasma sintering. For instance, sintering densification was promoted for microwave sintering of stainless steel powder as the particle size decreases [74], whereas decreasing the particle size causes sinterability to deteriorate for the case of microwave sintered aluminum [75]. This discrepancy in the observation stems from differences in microwave absorption of materials. It demonstrated that the ratio of particle size to penetration depth (skin depth), that is, the distance from the surface subjected to the electromagnetic waves to the location where the strength of the electric field is 36.8% of its magnitude at the surface, is a critical factor in governing the energy absorption, and it is more crucial than particle size for microwave sintering of conductive materials [75]. As a result of different efficiencies of converting electromagnetic energy to heat, powder with similar particle size represents a variety of sinterability behavior in the microwave furnace as opposed to the conventional furnaces. In the case of spark plasma sintering, an investigation into the influence of particle size on densification of copper powder demonstrated that decreasing particle size caused neck formation and densification to reduce when the sintering temperature was below 700 °C [76]. This is because the current flow was highly decreased for the finer particle size. Nevertheless, for the same copper powder when it was sintered above 700 °C, neck growth was promoted for the finer particle size since the number of contact points between particles per unit volume increased [77]. This is because the mass transport phenomena play a major role in sintering of finer particles.

3.2.4 Powder Segregation

In order to accomplish a reproducible part with homogeneity in terms of strength, density, and composition throughout the entire product, a random mixture of particles must be produced along each layer being spread by the roller. However, segregation, that is, the propensity of particulate mixtures to segregate or de-mix into various zones, inevitably occurs in particulate materials and leads to non-uniform spatial distributions. Segregation stems from the difference in properties of individual particles under the action of forces on a particulate mixture. The particle properties along with effects of different forces leading to segregation are summarized in the light of their scale and magnitude in Table 3.1.

Table 3.1 The properties of particles and the influence of forces acting on each property which lead to segregation (adapted from [78]).

Particle properties	Gravity force	Contact friction force	Electro-static force	Adhesion/van der Waals' forces	Viscous drag force	Inertial effects	Contact pressure
Size	None	None	Inverse	Inverse	Large	Some extent	Some extent
Shape	None	Some extent	Some extent	Large	Large	None	Some extent
Mass	Large	Large	Some extent	Some extent	Large	Large	Large
Density	Large	None	None	None	Large	Large	Some extent
Velocity	Some extent	Large	Large	None	Large	None	None
Acceleration	Large	Some extent	Some extent	None	Large	Large	Large
Composition	Large	Large	Large	Large	Some extent	Large	Some extent

From the literature, ten to thirteen mechanisms or patterns have been identified for segregation [78-81]. Among them, the most applicable ones to powder bed AM techniques are trajectory, impact, sieving, percolation, push-away, agglomeration and rolling segregation. Trajectory segregation appears in the circumstances where a cloud of particles is in motion. This segregation pattern is attributed to the difference in body forces such as gravity and the air drag force for small and large particles. The body forces are in proportion to the volume or the mass of particles while the air drag is correlated with particles' diameter. Accordingly, the dominating force for small particles is the air drag, whereas the body forces are dominated for large particles because the third power of diameter is proportional to the volume of particles. Thus, particles follow different paths based on their size when a cloud of particles is in motion. While particles are in motion, the impact effect is another mechanism which causes segregation to take place as a result of interparticle impact or impacts between particles and various surfaces of apparatus. After being stopped, moving particles will be found in various places due to the differences in inertia, resilience and other dynamic features for light and dense, small and large or elastic and non-elastic particles. Sieving segregation occurs when smaller particles percolate downwards through the void spaces between larger particles during a sliding process such as a shear motion. This situation causes finer particles to be collected at the bottom part of flowing layer as illustrated in Fig. 3.5a. Percolation mechanism is similar to the sieving mechanism. In the case of the percolation, larger particles are quite stationary and smaller ones percolate down by gravity while both smaller and larger particles are in movement in sieving mechanism. Whenever a mixture containing particles of distinct size can be rearranged, it is likely for the percolation to occur. With the existence of particles with different densities, the push-away mechanism is most likely to arise as can be seen in Fig. 3.5b in which the heavier particle at the top tends to push the lighter underneath aside. Agglomeration segregation is activated if particles tend to form agglomerates. This situation is the case for a multi-components mixture in which one component tend to agglomerate more readily compared to the others [78]. For example, the presence of moisture aids in the formation of agglomerates, especially for finer particles. As shown in Fig. 3.5c, the rolling mechanism contributes to segregation when a mixture containing coarse and fine particles fill onto a heap. The friction against rolling for larger particles is lower than that of smaller ones. Hence, larger particles proceed down the slope and accumulate near the edge of the heap. Rolling mechanism contributes more actively in segregation with increasing sphericity of particles.

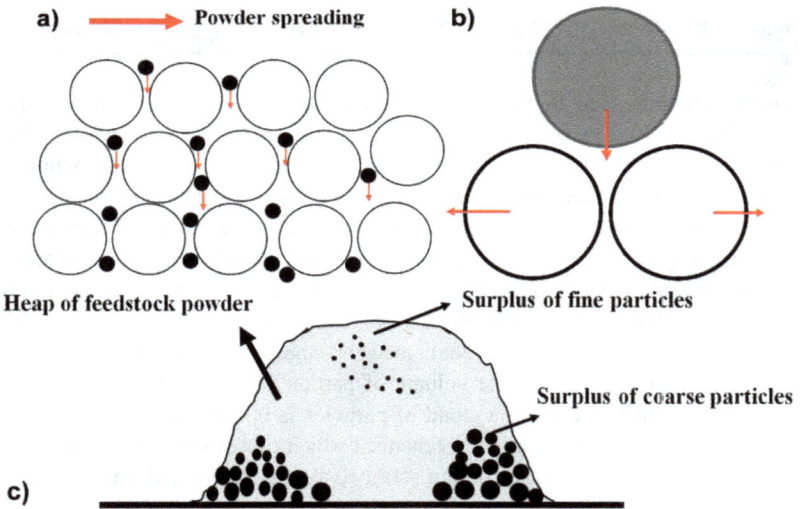

Figure 3.5 Schematic illustration of segregation mechanisms, a) sieving segregation, b) push-away, and c) rolling.

Segregation is likely to occur in different steps of the powder based AM techniques, including:

(i)   Feedstock preparation.

(ii)   Feeding feedstock into the feedstock platform.

(iii)   Spreading feedstock to the build platform via the roller.

Thus, understanding the factors influencing the segregation is of significant importance to minimize segregation. Influential factors on segregation broadly belong to four areas, namely:

(i)   materials related,

(ii)   operational or device related

(iii)   handling, and

(iv)   environmental related parameters [79].

The main material related factors are particle size, PSD, ratio of fine to coarse particles, particle shape, particle density, and particle morphology [79]. There is a general agreement that particle size is the most influential parameter. The coarser the particles, the greater the segregation will be [78, 82]. Segregation is more likely to take place for broader PSD [83, 84]. If the ratio of fine to coarse particles is enough to completely fill the voids between the coarse particles, less segregation is expected [78]. A mixture of spherically-shaped particles is more readily segregated than that of irregularly shaped particles due to its better flowability [85]. The main operational related factors can be the layer thickness, spreading rate, roller diameter, and its surface finish.

Segregation is also a very important factor in the production of large parts or series production using the powder bed AM techniques. To the best of the authors' knowledge, very few studies in this area have been conducted in the AM domain [86, 87]. However, numerous attempts have been made by researchers from engineering, pharmaceutical and food industries using experimental and/or simulation-based methodologies since segregation has ubiquitous presence wherever particulate systems are involved [88, 89]. Results of these studies would be beneficial in the investigation of segregation for powder based AM techniques. With this regards, evaluating the segregation behavior of powder using the rotating drum described under powder flowability section (3.2.1) can give invaluable information on the powder segregation for AM area [86].

3.3 Ink Selection

The selection of ink, either binder or solvent, is a crucial factor and is carried out concurrently with the selection of printing methods. Regardless of a selected printing method, the ink deposited by print heads must penetrate into the powder bed and selectively bind feedstock in a way that the powder bed arrangement will remain unaffected after a subsequent layer being spread. Furthermore, after drying or curing of the ink, the regions demarcated by the print head must have sufficient strength to be removed from the surrounding loose powder and handled gently while part edges are retained intact. Ink is of prime importance in inkjet 3DP as its interaction with the feedstock binds particles to each other, avoiding layer misalignment or shifting in a course of build cycle. Furthermore, sufficient binding strength enables safe handling of the as-built parts and fabrication of thin walled structures with high aspect ratio.

In order to reliably deposit ink that effectively interacts with the powder bed, the ink must satisfy certain rheological criteria in terms of surface tension and viscosity [90-92]. Several additive components might be added to ink to aid the printing process and/or affect final properties of the printed part. These additives include, water, surfactants, organic solvents and co-solvents, buffers, biocides, sequestering agent, viscosity modifiers, low molecular weight polymers, etc. [91-94]. Polyvinylpyrrolidone, for example, at the concentration of 0.05% causes better powder flowability and enhances contact adhesion of particles [95]. Interaction of powder and binder materials is a complex phenomenon and efforts should be made to develop a suitable feedstock-ink combination [96, 97]. The most significant parameters affecting the interaction between powder and ink's droplets are:

(i)   Droplets spreading behavior.

(ii)   Powder-droplets wettability.

(iii)   Reactivity [98].

The wettability is governed by many factors, including:

(i)   The topographic characteristics of the powder bed's surface and individual particles.

(ii)   The chemical composition of ink and its viscosity.

(iii)   The contact angle between ink and the powder bed.

(iv)   Possible chemical reaction taking place between ink and the powder bed [99].

When wettability and reaction time are too low, powder bed rearrangement will detrimentally affect the inkjet 3DP, whereas too high wetting worsens resolution and dimensional accuracy [40, 42]. The residue after debinding is another important issue that must be considered due to its effect on purity and strength of final parts. Thermal analysis techniques such as thermogravimetric analysis can be utilized to examine a decomposition temperature and an amount of a residue for a given thermal cycle. The residue ranges from being nonexistent to substantially contributing to the final strength of the inkjet 3D printed pars. Salts used in the salt solutions jetting on powder bed method are the examples for contributing the binder residue in ultimate properties of the printed parts. In the case of the other printing methods, binders with no residue are the ideal choice.

In general, binder materials can be classified into eight different classes according to Utela et al. [100], only three of which have been used in metal inkjet 3DP:

(i)   standard water-based binders such as commercially available ones [101],

(ii)   polymer solution binders such as PVA which is used in conjunction with solvent to create binding [102], and

(iii)   metal salt solutions [103].

The standard or organic binders bind a broad range of powder. These binders are fugitive and leave little or no ash residue once thermally decomposed during the debinding process. Aqueous metal salt solutions leave behind reduced metal residue after heat treating of the as-build parts, contributing to the part's final properties [104].

Summary

This chapter suggests that based on research conducted so far four factors namely, powder flowability, powder packing density, powder sinterability, and powder segregation are critical for the material selection. Adequate flowability which enables homogenous spreading of a thin layer of powder can be examined under static or dynamic conditions by several test methods. Both powder flowability and packing density are significantly related to interparticle forces which are governed by the powder feedstock characteristics in terms of particle size, particle size distribution, particle shape etc. In addition to the feedstock features, powder packing density is associated with build cycle's parameters. In general, powder with spherical shape and particle size larger than 20 μm provides good flowability alongside the packing density. Powder sinterability is also linked with the characteristics of feedstock. The influence of some feedstock characteristics on sinterability contrasts with powder flowability and the packing density. Through a variety of mechanisms, powder segregation, which is actively involved in any particulate systems, influences feedstock preparation as well as build cycle. However, the effects of segregation have not been critically investigated in the AM sector. Comprehending the impact of the feedstock features in altering each factor, feedstock materials can be selected to maximize powder flowability, packing density, and sinterability while powder segregation should be minimized. A set of criteria for ink selection is required to achieve successful build cycle as well as obtain sufficient handling strength in an as-build part. These criteria include rheological properties and wettability. The residue after the debinding process is another important consideration when it comes to the ink selection.

References

1. Sachs, E.M., et al., Jetting layers of powder and the formation of fine powder beds thereby. 2003, Google Patents.

2. Moon, J., et al., Slurry Chemistry Control to Produce Easily Redispersible Ceramic Powder Compacts. Journal of the American Ceramic Society, 2000. 83(10): p. 2401-2408. https://doi.org/10.1111/j.1151-2916.2000.tb01568.x

3. Turker, M., D. Godlinski, and F. Petzoldt, Effect of production parameters on the properties of IN 718 superalloy by three-dimensional printing. Materials Characterization, 2008. 59(12): p. 1728-1735. https://doi.org/10.1016/j.matchar.2008.03.017

4. Cima, M., et al., Three-dimensional printing techniques. 1995, Google Patents.

5. Cima, L.G. and M.J. Cima, Preparation of medical devices by solid free-form fabrication methods. 1996, Google Patents.

6. ASTM F3049-14, S.G.f.C.P.o.M.P.U.f.A.M.P., ASTM International, West Conshohocken, PA, 2014, www.astm.org.

7. ISO 17296-3:2014, A.m.-.-G.p.-.-P.M.c.a.c.t.m., ISO International Standards.

8. Ziegelmeier, S., et al., An experimental study into the effects of bulk and flow behaviour of laser sintering polymer powders on resulting part properties. Journal of Materials Processing Technology, 2015. 215: p. 239-250. https://doi.org/10.1016/j.jmatprotec.2014.07.029

9. Goodridge, R.D., C.J. Tuck, and R.J.M. Hague, Laser sintering of polyamides and other polymers. Progress in Materials Science, 2012. 57(2): p. 229-267. https://doi.org/10.1016/j.pmatsci.2011.04.001

10. Evans, R.S., et al. SLS materials development method for rapid manufacturing. in 16th Solid Freeform Fabrication Symposium, SFF 2005. 2005.

11. Prescott, J.K. and R.A. Barnum, On powder flowability. Pharmaceutical Technology, 2000. 24(10): p. 60-84+236.

12. Ploof, D.A. and J.W. Carson, Quality control tester to measure relative flowability of powders. Bulk Solids Handling, 1994. 14(1): p. 127-132.

13. Schulze, D., Measuring powder flowability: A comparison of test methods. Part I. Powder and Bulk Engineering, 1996. 10(4): p. 45-61.

14. Schwedes, J., Review on testers for measuring flow properties of bulk solids. Granular Matter, 2003. 5(1): p. 1-43. https://doi.org/10.1007/s10035-002-0124-4

15. Krantz, M., H. Zhang, and J. Zhu, Characterization of powder flow: Static and dynamic testing. Powder Technology, 2009. 194(3): p. 239-245. https://doi.org/10.1016/j.powtec.2009.05.001

16. Schulze, D., Powders and bulk solids: Behavior, characterization, storage and flow. First ed. Powders and Bulk Solids: Behavior, Characterization, Storage and Flow. 2008: Springer-Verlag Berlin Heidelberg. 1-511.

17. Standards, I.I., ISO 4324:1977, Surface active agents -- Powders and granules -- Measurement of the angle of repose. 1977, ISO International Standards.

18. Schmid, M., et al. Flowability of powders for Selective Laser Sintering (SLS) investigated by Round Robin Test. in High Value Manufacturing: Advanced Research in Virtual and Rapid Prototyping - Proceedings of the 6th International Conference on Advanced Research and Rapid Prototyping, VR@P 2013. 2014.

19. http://www.mercuryscientific.com/

20. Amado, A., et al. Advances in SLS powder characterization. in 22nd Annual International Solid Freeform Fabrication Symposium - An Additive Manufacturing Conference, SFF 2011. 2011.

21. Ziegelmeier, S., et al., An experimental study into the effects of bulk and flow behaviour of laser sintering polymer powders on resulting part properties. Journal of Materials Processing Technology, 2015. 215(1): p. 239-250. https://doi.org/10.1016/j.jmatprotec.2014.07.029

22. Van den Eynde, M., L. Verbelen, and P. Van Puyvelde, Assessing polymer powder flow for the application of laser sintering. Powder Technology, 2015. 286: p. 151-155. https://doi.org/10.1016/j.powtec.2015.08.004

23. Spierings, A.B., et al., Powder flowability characterisation methodology for powder-bed-based metal additive manufacturing. Progress in Additive Manufacturing, 2016. 1(1): p. 9-20. https://doi.org/10.1007/s40964-015-0001-4

24. Castellanos, A., The relationship between attractive interparticle forces and bulk behaviour in dry and uncharged fine powders. Advances in Physics, 2005. 54(4): p. 263-376. https://doi.org/10.1080/17461390500402657

25. Thomas, Y., et al. Effect of atmospheric humidity and temperature on the flowability of lubricated powder metallurgy mixes. in Advances in Powder Metallurgy and Particulate Materials - 2009, Proceedings of the 2009 International Conference on Powder Metallurgy and Particulate Materials, PowderMet 2009. 2009.

26. Spath, S. and H. Seitz, Influence of grain size and grain-size distribution on workability of granules with 3D printing. International Journal of Advanced Manufacturing Technology, 2014. 70(1-4): p. 135-144. https://doi.org/10.1007/s00170-013-5210-8

27. Jallo, L.J., et al., Prediction of Inter-particle Adhesion Force from Surface Energy and Surface Roughness. Journal of Adhesion Science and Technology, 2011. 25(4-5): p. 367-384. https://doi.org/10.1016/S0032-5910(02)00046-3

28. Forsyth, A.J., S. Hutton, and M.J. Rhodes, Effect of cohesive interparticle force on the flow characteristics of granular material. Powder Technology, 2002. 126(2): p. 150-154. https://doi.org/10.1016/S0032-5910(02)00046-3

29. Visser, J., Van der Waals and other cohesive forces affecting powder fluidization. Powder Technology, 1989. 58(1): p. 1-10.

30. Schulze, D., Powders and Bulk Solids-Behavior, Characterization, Storage and Flow. 2008: Springer-Verlag Berlin Heidelberg. XVI, 512.

31. Chan, L.C.Y. and N.W. Page, Particle fractal and load effects on internal friction in powders. Powder Technology, 1997. 90(3): p. 259-266. https://doi.org/10.1016/S0032-5910(96)03228-7

32. Freeman, R., Measuring the flow properties of consolidated, conditioned and aerated powders - A comparative study using a powder rheometer and a rotational shear cell. Powder Technology, 2007. 174(1-2): p. 25-33. https://doi.org/10.1016/j.powtec.2006.10.016

33. Faqih, A.M.N., et al., Effect of moisture and magnesium stearate concentration on flow properties of cohesive granular materials. International Journal of

Pharmaceutics, 2007. 336(2): p. 338-345.
https://doi.org/10.1016/j.ijpharm.2006.12.024

34. Zhu, J. and H. Zhang, Fluidization additives to fine powders. 2004, Google Patents.

35. Verbelen, L., et al., Characterization of polyamide powders for determination of laser sintering processability. European Polymer Journal, 2016. 75: p. 163-174. https://doi.org/10.1016/j.eurpolymj.2015.12.014

36. Butscher, A., et al., Printability of calcium phosphate powders for three-dimensional printing of tissue engineering scaffolds. Acta Biomaterialia, 2012. 8(1): p. 373-385. https://doi.org/10.1016/j.actbio.2011.08.027

37. Seville, J., U. Tüzün, and R. Clift, Processing of Particulate Solids. 1997, Dordrecht: Springer Netherlands. 372. https://doi.org/10.1007/978-94-009-1459-9

38. Zhu, H.H., J.Y.H. Fuh, and L. Lu, The influence of powder apparent density on the density in direct laser-sintered metallic parts. International Journal of Machine Tools and Manufacture, 2007. 47(2): p. 294-298. https://doi.org/10.1016/j.ijmachtools.2006.03.019

39. Spath, S., P. Drescher, and H. Seitz, Impact of Particle Size of Ceramic Granule Blends on Mechanical Strength and Porosity of 3D Printed Scaffolds. Materials, 2015. 8(8): p. 4720. https://doi.org/10.3390/ma8084720

40. Zhou, Z., et al., Printability of calcium phosphate: Calcium sulfate powders for the application of tissue engineered bone scaffolds using the 3D printing technique. Materials Science and Engineering: C, 2014. 38: p. 1-10. https://doi.org/10.1016/j.msec.2014.01.027

41. Kojima, T. and J.A. Elliott, Incipient flow properties of two-component fine powder systems and their relationships with bulk density and particle contacts. Powder Technology, 2012. 228: p. 359-370. https://doi.org/10.1016/j.powtec.2012.05.052

42. Lanzetta, M. and E. Sachs, Improved surface finish in 3D printing using bimodal powder distribution. Rapid Prototyping Journal, 2003. 9(3): p. 157-166. https://doi.org/10.1108/13552540310477463

43. Zou, R.-P., et al., Packing of Cylindrical Particles with a Length Distribution. Journal of the American Ceramic Society, 1997. 80(3): p. 646-652. https://doi.org/10.1111/j.1151-2916.1997.tb02880.x

44. Yu, A.B. and N. Standish, Characterisation of non-spherical particles from their packing behaviour. Powder Technology, 1993. 74(3): p. 205-213. https://doi.org/10.1016/0032-5910(93)85029-9

45. Kyrylyuk, A.V. and A.P. Philipse, Effect of particle shape on the random packing density of amorphous solids. physica status solidi (a), 2011. 208(10): p. 2299-2302. https://doi.org/10.1002/pssa.201000361

46. Abdullah, E.C. and D. Geldart, The use of bulk density measurements as flowability indicators. Powder Technology, 1999. 102(2): p. 151-165. https://doi.org/10.1016/S0032-5910(98)00208-3

47. Mostafaei, A., et al., Microstructural evolution and mechanical properties of differently heat-treated binder jet printed samples from gas- and water-atomized alloy 625 powders. Acta Materialia, 2017. 124: p. 280-289. https://doi.org/10.1016/j.actamat.2016.11.021

48. Shanjani, Y., E. Toyserkani, and C. Wei. Modeling and characterization of biomaterials spreading properties in powder-based rapid prototyping techniques. in ASME International Mechanical Engineering Congress and Exposition, Proceedings. 2008.

49. Shanjani, Y. and E. Toyserkani. Material spreading and compaction in powder-based solid freeform fabrication methods: Mathematical modeling. in 19th Annual International Solid Freeform Fabrication Symposium, SFF 2008. 2008.

50. ASTM B243-16, S.T.o.P.M., ASTM International, West Conshohocken, PA, 2016, www.astm.org

51. Dourandish, M., D. Godlinski, and A. Simchi, 3D printing of biocompatible PM-materials, in Materials Science Forum. 2007. p. 453-456.

52. Chappell, J.S., T.A. Ring, and J.D. Birchall, Particle size distribution effects on sintering rates. Journal of Applied Physics, 1986. 60(1): p. 383-391. https://doi.org/10.1063/1.337659

53. Liu, Z.Y., T.B. Sercombe, and G.B. Schaffer, The effect of particle shape on the sintering of aluminum. Metallurgical and Materials Transactions A: Physical Metallurgy and Materials Science, 2007. 38(6): p. 1351-1357. https://doi.org/10.1007/s11661-007-9153-2

54. Park, D.Y., et al., Effects of Particle Sizes on Sintering Behavior of 316L Stainless Steel Powder. Metallurgical and Materials Transactions A, 2013. 44(3): p. 1508-1518. https://doi.org/10.1007/s11661-012-1477-x

55.   Mamen, B., et al., Experimental and numerical analysis of the particle size effect on the densification behaviour of metal injection moulded tungsten parts during sintering. Powder Technology, 2015. 270, Part A: p. 230-243.

56.   Nandwana, P., et al., Powder bed binder jet 3D printing of Inconel 718: Densification, microstructural evolution and challenges. Current Opinion in Solid State and Materials Science, 2017. https://doi.org/10.1016/j.cossms.2016.12.002

57.   Wautelet, M., On the shape dependence of the melting temperature of small particles. Physics Letters A, 1998. 246(3): p. 341-342. https://doi.org/10.1016/S0375-9601(98)00538-6

58.   Samsonov, V.M., S.A. Vasilyev, and A.G. Bembel, Size dependence of the melting temperature of metallic nanoclusters from the viewpoint of the thermodynamic theory of similarity. The Physics of Metals and Metallography, 2016. 117(8): p. 749-755. https://doi.org/10.1134/S0031918X16080135

59.   Allen, G.L., et al., Small particle melting of pure metals. Thin Solid Films, 1986. 144(2): p. 297-308. https://doi.org/10.1016/0040-6090(86)90422-0

60.   Sakai, H., Surface-induced melting of small particles. Surface Science, 1996. 351(1-3): p. 285-291. https://doi.org/10.1016/0039-6028(95)01263-X

61.   Ting, J.M. and R.Y. Lin, Effect of particle size distribution on sintering - Part II Sintering of alumina. Journal of Materials Science, 1995. 30(9): p. 2382-2389. https://doi.org/10.1007/BF01184590

62.   Ting, J.M. and R.Y. Lin, Effect of particle-size distribution on sintering - Part I Modelling. Journal of Materials Science, 1994. 29(7): p. 1867-1872. https://doi.org/10.1007/BF00351306

63.   Bjørk, R., et al., The effect of particle size distributions on the microstructural evolution during sintering. Journal of the American Ceramic Society, 2013. 96(1): p. 103-110. https://doi.org/10.1111/jace.12100

64.   Lim, L.C., P.M. Wong, and J. Ma, Colloidal processing of sub-micron alumina powder compacts. Journal of Materials Processing Technology, 1997. 67(1): p. 137-142. https://doi.org/10.1016/S0924-0136(96)02833-6

65.   Patterson, B.R. and J.A. Griffin. EFFECT OF PARTICLE SIZE DISTRIBUTION ON SINTERING OF TUNGSTEN. in Modern Developments in Powder Metallurgy. 1985.

66.   Shiau, F.S., T.T. Fang, and T.H. Leu, Effect of particle-size distribution on the microstructural evolution in the intermediate stage of sintering. Journal of the

American Ceramic Society, 1997. 80(2): p. 286-290.
https://doi.org/10.1111/j.1151-2916.1997.tb02828.x

67. Hay, R.A., W.C. Moffatt, and H. Kent Bowen, Sintering behavior of uniform-sized
     α-Al2O3 powder. Materials Science and Engineering A, 1989. 108(C): p. 213-219.
     https://doi.org/10.1016/0921-5093(89)90422-X

68. Sordelet, D.J. and M. Akinc, Sintering of Monosized, Spherical Yttria Powders.
     Journal of the American Ceramic Society, 1988. 71(12): p. 1148-1153.
     https://doi.org/10.1111/j.1151-2916.1988.tb05807.x

69. Yan, M.F., Microstructural control in the processing of electronic ceramics.
     Materials Science and Engineering, 1981. 48(1): p. 53-72.
     https://doi.org/10.1016/0025-5416(81)90066-5

70. Liniger, E.G. and R. Raj, Spatial Variations in the Sintering Rate of Ordered and
     Disordered Particle Structures. Journal of the American Ceramic Society, 1988.
     71(9): p. C-408-C-410. https://doi.org/10.1111/j.1151-2916.1988.tb06423.x

71. Ma, J. and L.C. Lim, Effect of particle size distribution on sintering of
     agglomerate-free submicron alumina powder compacts. Journal of the European
     Ceramic Society, 2002. 22(13): p. 2197-2208. https://doi.org/10.1016/S0955-
     2219(02)00009-2

72. Taruta, S., et al., Influence of coarse particle shape on packing and sintering of
     bimodal size-distributed alumina powder mixtures. Journal of Materials Science
     Letters, 1993. 12(6): p. 424-426. https://doi.org/10.1007/BF00609173

73. Cutler, I.B. and R.E. Henrichsen, Effect of Particle Shape on the Kinetics of
     Sintering of Glass. Journal of the American Ceramic Society, 1968. 51(10): p.
     604-604. https://doi.org/10.1111/j.1151-2916.1968.tb13334.x

74. Ertugrul, O., et al., Effect of particle size and heating rate in microwave sintering
     of 316L stainless steel. Powder Technology, 2014. 253: p. 703-709.
     https://doi.org/10.1016/j.powtec.2013.12.043

75. Crane, C.A., et al., The effects of particle size on microwave heating of metal and
     metal oxide powders. Powder Technology, 2014. 256: p. 113-117.
     https://doi.org/10.1016/j.powtec.2014.02.008

76. Diouf, S., C. Menapace, and A. Molinari, Study of effect of particle size on
     densification of copper during spark plasma sintering. Powder Metallurgy, 2012.
     55(3): p. 228-234. https://doi.org/10.1179/1743290111Y.0000000019

77.   Diouf, S. and A. Molinari, Densification mechanisms in spark plasma sintering: Effect of particle size and pressure. Powder Technology, 2012. 221: p. 220-227. https://doi.org/10.1016/j.powtec.2012.01.005

78.   Levy, A. and C.J. Kalman, Handbook of Conveying and Handling of Particulate Solids. 2001: Elsevier Science.

79.   Tang, P. and V.M. Puri, Methods for Minimizing Segregation: A Review. Particulate Science and Technology, 2004. 22(4): p. 321-337. https://doi.org/10.1080/02726350490501420

80.   Mosby, J., S.R. de Silva, and G.G. Enstad, Segregation of particulate materials – Mechanisms and testers. KONA Powder and Particle Journal, 1996. 14(May): p. 31-43. https://doi.org/10.14356/kona.1996008

81.   Williams, J.C., The segregation of particulate materials. A review. Powder Technology, 1976. 15(2): p. 245-251. https://doi.org/10.1016/0032-5910(76)80053-8

82.   Rhodes, M., Mixing and Segregation, in Introduction to Particle Technology. 2008, John Wiley & Sons, Ltd. p. 293-310. https://doi.org/10.1002/9780470727102.ch11

83.   Thomson, F.M., Storage and Flow of Particulate Solids in Handbook of Powder Science and Technology, M. Fayed and L. Otten, Editors. 1997, Springer US: New York: Chapman and Hall. . p. 389-486.

84.   Bates, L., G.D. Hayes, and B.M.H. Board, User Guide to Segregation. 1997: British Materials Handling Board.

85.   Massol-Chaudeur, S., H. Berthiaux, and J.A. Dodds, Experimental study of the mixing kinetics of binary pharmaceutical powder mixtures in a laboratory hoop mixer. Chemical Engineering Science, 2002. 57(19): p. 4053-4065. https://doi.org/10.1016/S0009-2509(02)00262-2

86.   Klisiewicz, P., J.A. Roberts, and N.A. Pohlman, Segregation of titanium powder with polydisperse size distribution: Spectral and correlation analyses. Powder Technology, 2015. 272: p. 204-210. https://doi.org/10.1016/j.powtec.2014.11.029

87.   Haeri, S., et al., Discrete element simulation and experimental study of powder spreading process in additive manufacturing. Powder Technology, 2017. 306: p. 45-54. https://doi.org/10.1016/j.powtec.2016.11.002

88.   Schlick, C.P., et al., Modeling segregation of bidisperse granular materials using physical control parameters in the quasi-2D bounded heap. AIChE Journal, 2015. 61(5): p. 1524-1534. https://doi.org/10.1002/aic.14780

89.   Ely, D.R., Dry powder segregation and flowability: Experimental and numerical studies, in Industrial and Physical Pharmacy. 2010, Purdue University: Ann Arbor. p. 146.

90.   Utela, B., et al., A review of process development steps for new material systems in three dimensional printing (3DP). Journal of Manufacturing Processes, 2008. 10(2): p. 96-104. https://doi.org/10.1016/j.jmapro.2009.03.002

91.   The Chemistry of Inkjet Inks. 2011: WORLD SCIENTIFIC.

92.   Fink, J.K., The Chemistry of Printing Inks and Their Electronics and Medical Applications. 2014: John Wiley & Sons, Inc.

93.   Oriakhi, C. and T.M. Lambright, Calcium aluminate cement compositions for solid freeform fabrication. 2007, Google Patents.

94.   Fink, J.K., The Chemistry of Printing Inks and Their Electronics and Medical Applications. 2014, John Wiley & Sons, Inc. p. 1-361.

95.   Xiong, Y., C. Qian, and J. Sun, Fabrication of porous titanium implants by three-dimensional printing and sintering at different temperatures. Dental Materials Journal, 2012. 31(5): p. 815-820. https://doi.org/10.4012/dmj.2012-065

96.   Guvendiren, M., et al., Designing Biomaterials for 3D Printing. ACS Biomaterials Science & Engineering, 2016.

97.   Khalyfa, A., et al., Development of a new calcium phosphate powder-binder system for the 3D printing of patient specific implants. Journal of Materials Science: Materials in Medicine, 2007. 18(5): p. 909-916. https://doi.org/10.1007/s10856-006-0073-2

98.   McEleney, P., et al., Investigations on drop penetration and wetting characteristics of powder-liquid systems in relation to the mixing of acrylic bone cement. International Journal of Nano and Biomaterials, 2010. 3(1): p. 20-35. https://doi.org/10.1504/IJNBM.2010.036105

99.   Sachs, E., et al., Three-Dimensional Printing: The Physics and Implications of Additive Manufacturing. CIRP Annals - Manufacturing Technology, 1993. 42(1): p. 257-260. https://doi.org/10.1016/S0007-8506(07)62438-X

100. Utela, B.R., et al., Development Process for Custom Three-Dimensional Printing (3DP) Material Systems. Journal of Manufacturing Science and Engineering, 2010. 132(1): p. 011008-011008. https://doi.org/10.1115/1.4000713

101. Verlee, B., T. Dormal, and J. Lecomte-Beckers, Density and porosity control of sintered 316L stainless steel parts produced by additive manufacturing. Powder Metallurgy, 2012. 55(4): p. 260-267. https://doi.org/10.1179/0032589912Z.00000000082

102. Wiria, F.E., S. Maleksaeedi, and Z. He, Manufacturing and characterization of porous titanium components. Progress in Crystal Growth and Characterization of Materials, 2014. 60: p. 94-98. https://doi.org/10.1016/j.pcrysgrow.2014.09.001

103. Sachs, E.M., et al., Metal and ceramic containing parts produced from powder using binders derived from salt. 2003.

104. Hong, S.B., et al., Corrosion behavior of advanced titanium-based alloys made by three-dimensional printing (3DPTM) for biomedical applications. Corrosion Science, 2001. 43(9): p. 1781-1791. https://doi.org/10.1016/S0010-938X(00)00181-5

CHAPTER 4

Inkjet Printing and Post Printing

Abstract

This chapter aims to introduce basic fundamentals of the inkjet technology including common methods for ink droplet projection and the interaction between the generated drops and the powder bed. Printing related parameters such as ink saturation level, layer thickness, and build orientation are addressed and the effects of these parameters on the properties of the fabricated parts are carefully analysed. Lastly, post-printing parameters for transformation from the as-built parts into end parts, including removing the parts from the powder bed followed by debinding and sintering are discussed.

Keywords

Inkjet Technology, Printing Parameter, Print head, Saturation Level, Layer Thickness, Drying Time, Build Orientation, Debinding, Sintering, Infiltration, Roller Speed

4.1 Introduction

Many variables are involved in inkjet 3DP to obtain end products with desired properties. These variables can be classified into four major categories:

- (i)   design related parameters
  - a. smallest feature sizes
  - b. dimensional accuracy
  - c. smallest pore size
  - d. strut thickness in lattice
- (ii)   materials related parameters (characteristics of powder feedstock and ink)
- (iii)   printing-related parameters
  - a. layer thickness
  - b. saturation level
  - c. build orientation
  - d. feed powder to layer thickness ratio
  - e. ink droplet size
- (iv)   post-printing related parameters
  - a. curing

b. depowdering

c. debinding

d. sintering

e. infiltration

Materials related parameters were thoroughly described in chapter 3. Printing-related parameters together with post-printing related parameters are introduced in this chapter. Finally, an overview of metallic materials under research and their key performance indicators will be addressed in chapter 5.

4.2   Fundamental of the Inkjet Technology

This section provides a general overview of drop formation methods and the interaction of ink droplets and the powder bed. Understanding the principals of the inkjet technology is vital for a better understanding of different variables playing important parts to realize the desired end products. For detailed information on any topic cited references in this chapter can be referred to.

4.2.1   Methods of Drop Generation

Inkjet printing technology involves the generation of small ink drops and the precise deposition of the drops on the powder bed. The print head is the heart of the inkjet technology which aims to fire ink drops at a high frequency and in a controlled manner with the capability of replication in terms of droplet size. Several methods for the generation of ink drops have been invented. The most common classification of drop generation methods is shown in Fig. 4.1 [1]. Currently, drop-on-demand (DOD) and continuous inkjet (CIJ) modes are the two major operation mechanisms of print heads through which ink drops are generated [2]. In both cases, ink flows through a minute orifice, commonly known as a nozzle. CIJ is based on producing a continuous stream of drop while the flow is impulsive in DOD. Typical emitted drop volumes lie in the range from 0.5 to 500 picolitres (pl), corresponding to drop diameters from 10 to 100 μm as described in Table 4-1. Ejected droplets strike the powder bed with the speed in the range of 5-30 m/s, depending on the type of technology used [1].

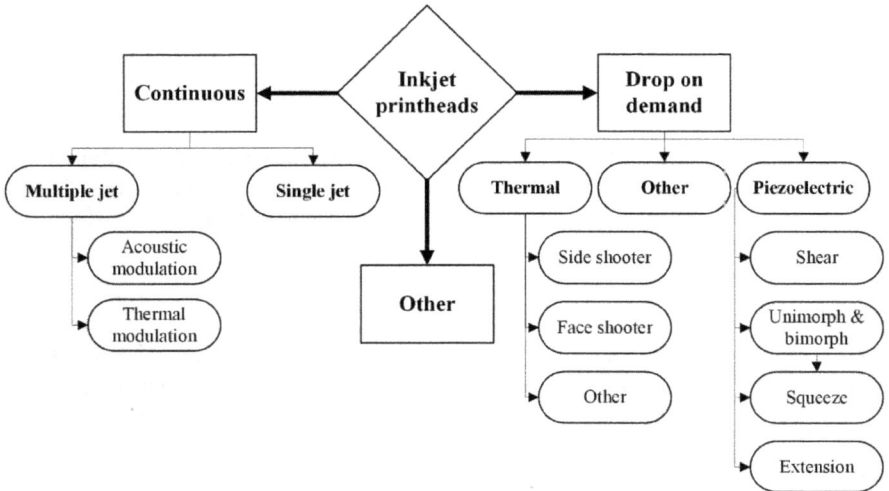

Figure 4.1 Classification of the inkjet printing technology [1].

Table 4-1 Typical drop volumes and equivalent drop diameters (1 pl=10^{-12} liter) [1].

Volume (pl)	Diameter (μm)
0.1	5.76
0.52	10
1	12.41
10	26.73
33.51	40
65.45	50
100	57.59
268.08	80
523.6	100
1000	124.07

4.2.1.1　　Continuous Inkjet (CIJ)

In the CIJ print head, a continuous stream of drops is produced. Fig. 4.2 displays the formation and positioning of ink droplets. As can be seen from the figure, when pressure is imposed on the liquid column behind the small nozzle, the liquid tends to break up into a series of drops in accordance with the Rayleigh instability [1-4]. Imparting a small charge to each ejected drop allows them to be directed and positioned using charged deflection plates while unwanted drops are being collected by a gutter. These unused drops may be recycled. Similar to the single-jet CIJ, a couple of CIJ nozzles can be operating in parallel to form multiple-jet CIJ in which several drop streams are generated from a single print head. In general, the diameter of the ejected drops is slightly larger than the nozzle diameter. CIJ systems generate drops with diameters typically >50 μm [1].

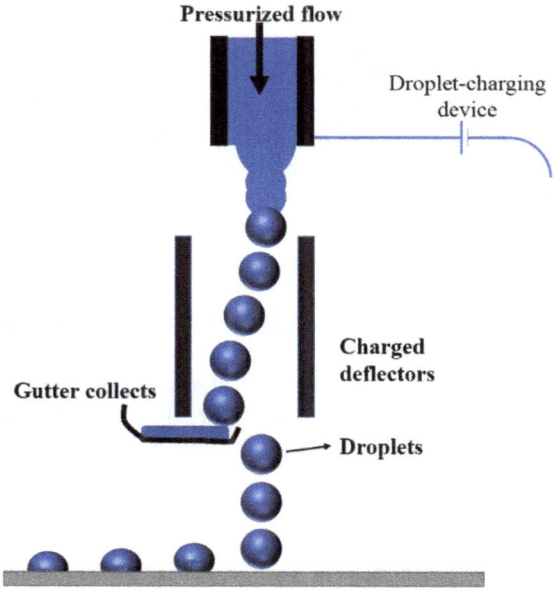

Figure 4.2　Schematic illustration showing the operation principles of CIJ print head (adapted from [3]).

4.2.1.2   Drop on Demand (DOD)

Familiar office and home inkjet printers operate based on DOD technologies to fire droplets. A DOD print head comprises of an ink reservoir with an ink inlet on one end that allows ink to be impelled in and out, while there are multiple nozzles (i.e. a few hundreds or thousands) on the other end through which droplets are projected [5]. In contrast to the CIJ print head, the DOD generates individual droplets when required and there is no drop deflection or selection system. Thus, drop positioning is attained by placing the print head close to the powder bed and above the desired location together with firing each individual nozzle only as required. Using an actuation mechanism, sufficient energy transfers to a suitable volume of ink that enables ejection of a drop with acoustic frequencies in the range from 1 to 20 kHz. Among different actuations mechanisms, piezoelectric actuation and thermal (or bubble) actuation are the most common techniques [6-9]. Accordingly, thermal and piezoelectric are the two types of DOD inkjet print heads.

In thermal inkjet (TIJ) print heads, also known as bubble-jet, an electrical current is passed through a small resistive heater (approximately 50 μm in diameter) which is in contact with the ink in the reservoir, causing superheating (about 300 °C) and vaporization of a very thin layer of the ink adjacent to the heater [2, 7]. This rapid heating results in the formation and expansion of a vapour bubble in the ink chamber, providing energy necessary for fluid displacement to push a drop through the nozzle. This process that occurs within approximately two micro-seconds is schematically illustrated in Fig 4.3a. When the drop is ejected and the heat pulse is switched off, the bubble rapidly collapses that pulls fresh charge of ink into the reservoir through the ink inlet. Afterwards, the process is ready to start again. Several studies have demonstrated that the thin layer of ink next to the heater return to ambient temperature within 10 micro-seconds and the bulk of ink is heated only by 4-10 °C [6, 10-12]. Therefore, it is perfectly possible to use a TIJ print head for heat-sensitive fluid such as living cells with the intend of bioprinting of tissues [9]. Major drawbacks of the TIJ print head are limited operation voltages of nozzles and availability of inks. Thus, TIJ requires inks with the formulations that fall within a specific range of surface tensions, viscosities, vapour pressures, and boiling points [7, 8]. Most water-based inks are able to satisfy these criteria.

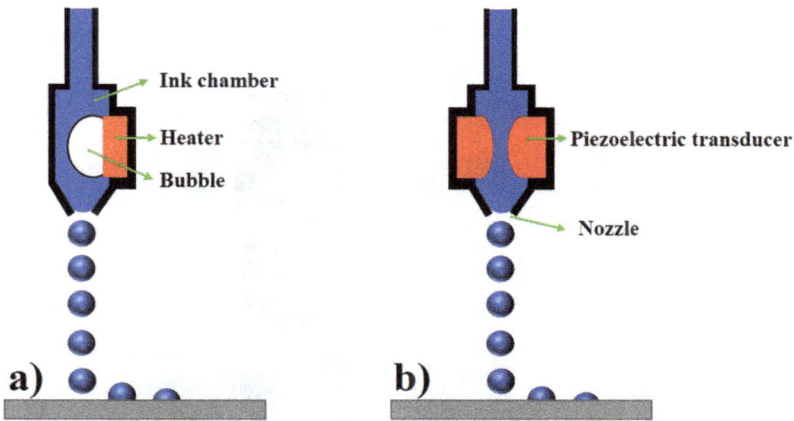

Figure 4.3. Schematic representation of the working principles of a DOD print head: a) thermal inkjet and b) piezoelectric inkjet (adapted from [3]).

The piezoelectric (Piezo) effect is the capability of some specific materials (e.g. quartz) to generate an electric charge when mechanical force is applied and vice versa. In Piezo inkjet print heads as depicted in Fig.4.3b, a voltage pulse drives a Piezo material next to the ink reservoir to generate a pressure in the reservoir, forcing a droplet of ink through the nozzle. Operation mechanism of Piezo print heads is essentially the same as that of TIJ but the pressure pulse is generated using mechanical actuation. In terms of technology, TIJ print heads are inferior to the Piezo ones since its [2, 7, 8, 13, 14]:

  i. lifespan is shorter (less solvent and acid resistant),
 ii. maximum speed is lower, and
iii. lower capability of controlling drop size and velocity.

Thus, new devices mostly use Piezo DOD technology although Piezo print heads are costly.

In almost all application of inkjet technology for 3DP, DOD is the method of choice due to its greater accuracy of placement and smaller droplet size as compared with CIJ print heads. Both TIJ and Piezo print heads are used in the inkjet 3DP domain as inks are mostly aqueous based solutions. Fig 4.4 shows a TIJ print head used in a Z Corp 310® machine which can jet either a solvent or a liquid binder. Fig 4.5 displays a ExOne Innovent Piezo print head.

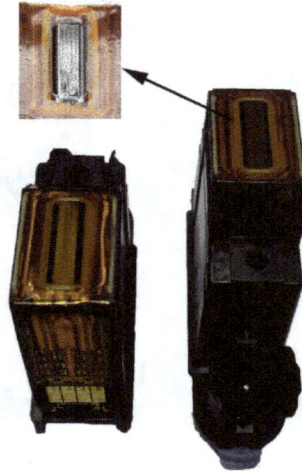

Figure 4.4. HP 10 print head with 304 nozzles and drop volumes of 35 pl which is utilized in Z Corp 310 machines for either solvent or binder jetting.

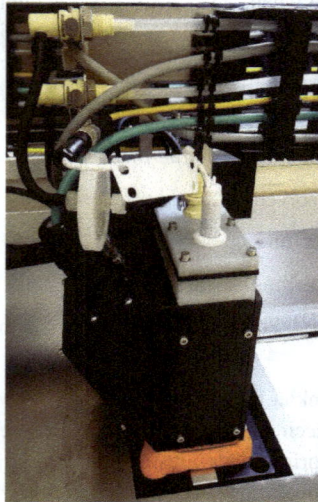

Figure 4.5. The Piezo print head of ExOne Innovent machine used for the liquid BJ on powder bed method.

4.2.2 Interaction between Ink Drop and Powder Bed

As mentioned above, the powder bed is struck by sphere like ink drops with diameter of tens of microns which are ejected from the print heads travelling at the velocity in the order of 10 m/s. Each layer of loosely spread powder bed can be considered as a semi-infinite porous medium comprising of fine solid particles. The impingement of ink droplet with such a media essentially differs from continuous media such as paper [3, 15]. Regardless of the media, a number of factors affect the impingement which include:

  i.   the rheological properties of ink,

  ii. velocity and mass of impinging drops, and

  iii. the nature of the media in terms of wettability.

The multiplicity of the factors has posed a challenge to the researchers to completely understand the mechanisms involved in the process [15-17]. Generally, when ink droplets contact with the powder bed, they do not retain their initial spherical shape and start to spread since their inertia results in a forced wetting.

Using high speed stroboscopic photography, Fan [16] investigated the complex interaction between ink droplets and alumina based powder bed in inkjet 3DP. A single droplet interaction with the powder bed is a two-stage process that consists of deformation/spreading followed by absorption. The kinetic energy of the initial droplet drives drop spreading across the powder bed until its energy has been consumed through dissipation or conversion into another form of energy such as surface energy. The interaction involves the dissipation of the kinetic energy followed by capillarity driven spreading and coalescence including intermingling of the droplet with the powder bed particles, penetration of the droplet-powder mixture into adjacent particles, and rearrangement of the powder particles. All these events take place rapidly and in a very short period of time. The increment in mass of the evolving particles aggregate is the dominating factor in the kinetic of the droplet-powder mixture, while the resistance of the powder bed to penetration diminishes the momentum of the impact penetration. Fig. 4.6 displays the evolved aggregates of particles with roughly spherical shape which form as result of the interaction of a single droplet ejected from a DOD print head with alumina powder bed. The spherical shaped aggregates are the favoured shape due to minimal surface energy after the capillarity driven rearrangement has occurred. These aggregates are the smallest feature size that can be created in a part fabricated using the inkjet 3DP technology. Accordingly, one way to reduce this finest feature is to reduce the feedstock particle size and ink drop size.

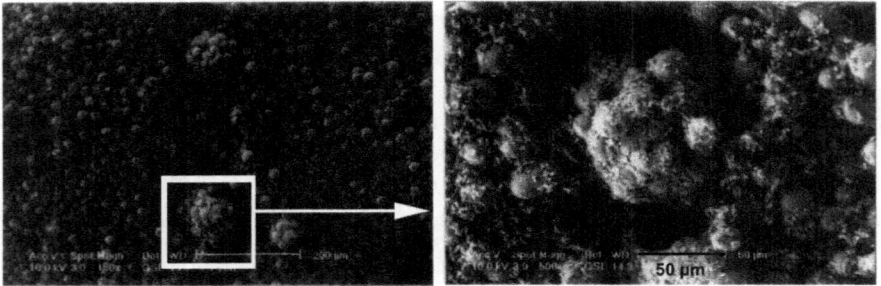

Figure 4.6 Top view SEM images of a single droplet interaction with the alumina powder bed. The droplet was fired by a DOD print head operating at the speed of 1.5 kHz (Drop size range:50-70 µm; Speed: 15±1.5 m/s) [17].

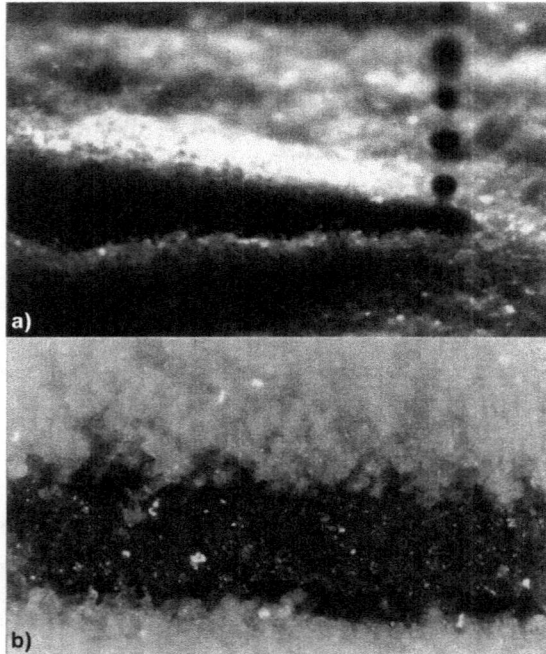

Figure. 4.7. a) High-speed frame of the impingements of a series of ink drops in a linear fashion with 30 µm platelet shaped alumina powder bed and b) the line created from the droplets projected by a CIJ print head with droplet size of 90 µm at a speed of 10 m/s [17].

Generating a train of drops with overlap in their placement makes it possible for the small aggregate to be extended into lines and plane. Accordingly, two-dimensional geometries representing the cross section of a part in accordance with the STL data can be achieved. The third dimension of the part is formed by accumulating these 2D slices vertically owing to penetration of ink to preceding layers. Fig. 4.7a shows projecting of a series of drops in a linear fashion which result in the formation of a line (Fig. 4.7b). The impact of ink droplets with powder bed is likely to eject the particles from the powder bed and form the trench as can be seen in Fig. 4.7a. Although this impact deformation ceases within 1 millisecond, the event leads to the formation of aggregates at the bottom of the crater which may penetrate to about tens of micrometers below the powder bed surface [16].

4.3 Printing Parameters

Various printing-related parameters need to be considered to obtain an inkjet 3D printed part with high dimensional resolution, accuracy, surface finish, integrity, and green strength. Major printing parameters include:

   i. Ink saturation level

  ii. Layer thickness

 iii. Drying time and power control

 iv. Build orientation

  v. Roller spread speed and feed-to-powder ratio

4.3.1 Ink Saturation Level

The ink saturation level (SL) is the ratio of the amount of air space between particles in each deposited layer which is occupied by deposited ink (V_{liquid}) to the total amount of air space (V_{air}). SL can be calculated using the following equations [18]:

$$SL = V_{liquid} / V_{air} \tag{2}$$

$$V_{air} = (1\text{-}PR) \times V_{solid} \tag{3}$$

Where, PR is powder packing density and V_{solid} is the volume of solid particles in a defined envelope. SL describes the amount of ink which will be deposited in each layer. Accordingly, a simple definition for SL could be the thickness of penetration when the ink is penetrating into a given layer of powder. SL might be as high as 170% for metal inkjet 3DP [19]. Software accompanied to an inkjet 3DP machine is able to distinguish two regions in each 2D section of a part, namely core and shell as displayed in Fig. 4.8.

Thus, two different values for SL can be applied to build the entire core and shell of a 3D part since each 2D section is composed of cores and shells. For instance, depositing greater amount of liquid on the shell regions makes it possible to stabilize the edges of a part while the lower amount is administrated in the interior area. Given the cores and shells SL, it is plausible to obtain enough stability in a part without over saturation or compromising the build time.

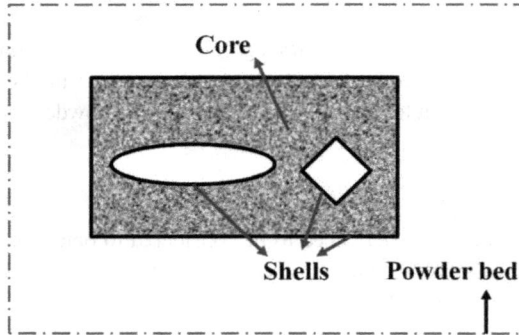

Figure 4.8 Two distinguishable regions by the software in a sliced 2D layer, the areas around the boundary of cross-section are the shells and the area inside the cross-sectional region filled with grey colour is the core.

SL is a crucial factor for the stability of the build cycle. Inadequate SL causes insufficient binding between adjacent particles within each layer, leading to the layer displacement during the build cycle. Furthermore, surface wear-out, delamination, and complete failure may occur during part removal from the powder bed. The SL should be high enough that the deposited ink sticks particles together across each layer as well as penetrates into the previous layer [20]. On the other hand, if the SL is too high, the capillarity driven diffusion of ink to the surrounding powder causes excess ink to wick away from the deposited regions so that some extra particles join to the part or even the roller. The lateral spreading of ink deteriorates the surface finish and dimensional accuracy of a part [19, 21]. Dimensional accuracy refers to the mismatch between the dimension of an as-build part and its designed specification. Moreover, higher SL increases the build time and may adversely affect the post-printing process. Thus, fine-tuning the amount of SL is crucial not only for the build cycle but also for minimizing cost of the inkjet 3DP technology. The influences of SL on dimensional accuracy and the corresponding pores interconnectivity of the as-build parts are represented in Fig. 4.9 and 4.10, respectively.

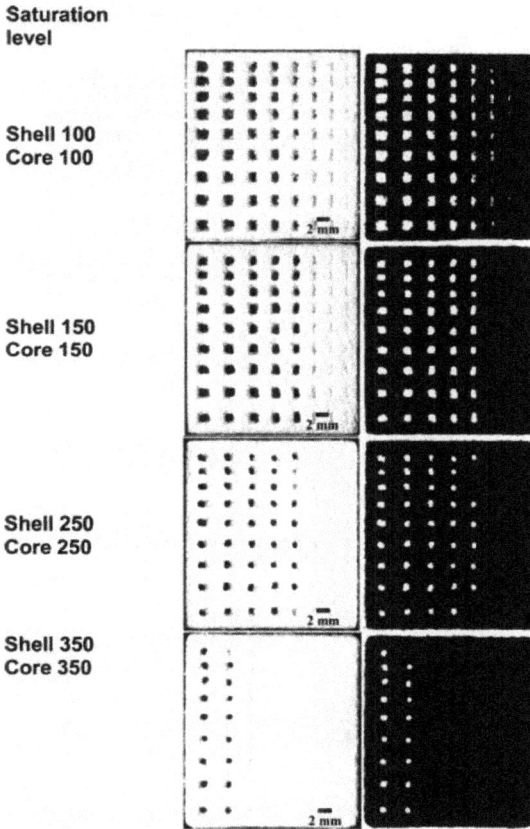

Figure 4.9 The effects of different values for SL on the dimensional accuracy of calcium phosphate based feedstock printed by the liquid BJ on powder bed method. The left side images are the digital scanned images of each samples (adapted from [22]).

As can be seen qualitatively and quantitatively from the figures, increasing the SL leads to significant reduction of dimensional accuracy and the interconnectivity. Activating binding agent in the powder feedstock (i.e. the SJ on powder bed approach) or providing a liquid binder (i.e. the liquid BJ on powder bed approach) is achieved by SL. Enhancements in both interlayer and intra-layer binding with increasing the value of SL per layer thickness enable the as-build parts to be successfully fabricated and then

73

removed from the loose surrounding powder. With further increasing SL, however, the amount of liquid bleeding from the existing interparticle spaces is increased which leads to poorer dimensional accuracy and surface finish. Fig 4.10 clearly demonstrates the adverse effect of increasing SL on the pores interconnectivity. Irrespective of the macro pores size, the interconnectivity is likely to deteriorate at higher SL. Furthermore, it is less plausible to create finer pores when SL is high.

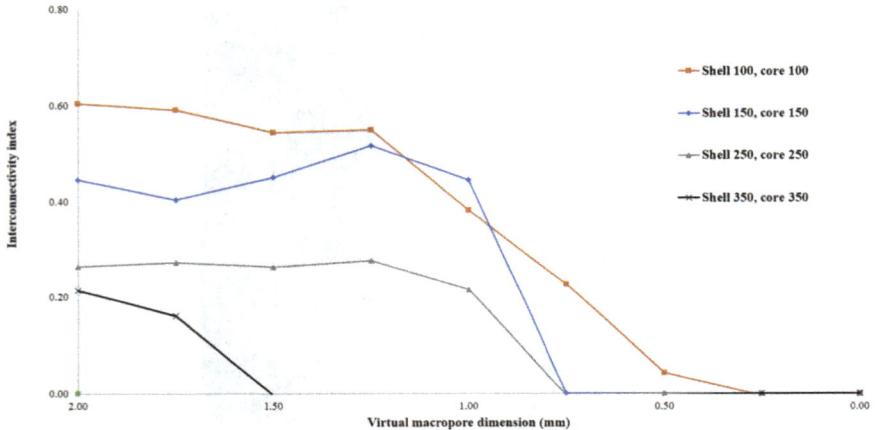

Figure 4.10 Variation of pores interconnectivity as a function of macropore dimensions for various values of SL. The interconnectivity index is the ratio between macropore area in the as-build parts and the area in the CAD file (adapted from [22]).

As it would be anticipated, increasing SL results enhance the green strength [22-24]. As mentioned in chapter 2, the green strength of the as-build parts refers to the initial strength after removing the parts from the powder bed and before conducting any post-printing processes. The green strength of a part results from the binding quality of the particles within each layer as well as neighbouring layers. Inadequate green strength leads to:

  i. Deterioration in dimensional accuracy

  ii. Poor surface finish

  iii. Distortion

  iv. Cracking, or failure of the inkjet 3D printed parts

Fig. 4.11 shows an increase in the green strength of plaster-based feedstock printed using the liquid BJ on powder bed method along with the amount of SL. In addition, using higher value of SL is more likely to improve consolidation of as-build parts and enhance the green density due to the vertical and lateral spreading of ink [22, 23]. Due to better green density, higher SL may promote the relative density of the sintered parts [22, 25].

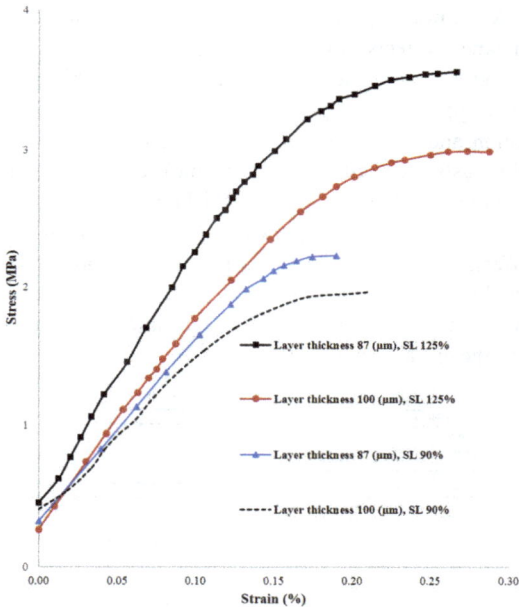

Figure 4.11 Stress-strain curves for as-build parts made of plaster-based ZP102 powder which were fabricated by the liquid BJ on powder bed method of Zb56 binder (adapted from [24]).

The desired value for SL depends on the powder packing density, ink-powder bed interaction, and layer thickness [19, 23, 24, 26]. The interaction between ink and powder bed is greatly associated with the intrinsic properties of powder feedstock and ink. Likewise, powder packing density is related to the intrinsic properties of the feedstock. Thus, from a practical standpoint, the layer thickness is the main variable that should be considered to figure out an appropriate value for SL.

4.3.2 Layer Thickness

Layer thickness refers to the height of each layer of the powder feedstock being spread by the roller to build a part layer-by-layer. The layer thickness must be larger than the largest particle so that the minimum value is governed by the largest particles within the particle size distribution of a given powder. The preferred layer thickness is three times of the particle size [27, 28]. Higher layer thickness generally ensures better powder spreading behavior (i.e. powder flowability) and shorter build time. However, as mentioned in chapter 3, layer thickness affects powder packing density since the roller compacts the feedstock while spreading it over the powder bed. In view of this, greater layer thickness leads to lower powder packing density [29]. Furthermore, a thicker layer causes poorer dimensional resolution and tolerance. Fig. 4.12 depicts a sphere sliced with a different layer thickness. Obviously, none of these layers thickness makes the exact curvature of the outside surface. However, the smaller layer thickness results in closer accord with the curvature. In general, a 100 μm layer thickness is used in metal inkjet 3DP. Fig. 4.13 shows profound effects of the layer thickness on porosity and mechanical properties of inkjet 3D printed and sintered Alumina samples. As can be seen, porosity percentage of the sintered samples constantly increases with layer thickness leading to a decrease in flexural strength and the elastic modulus [30].

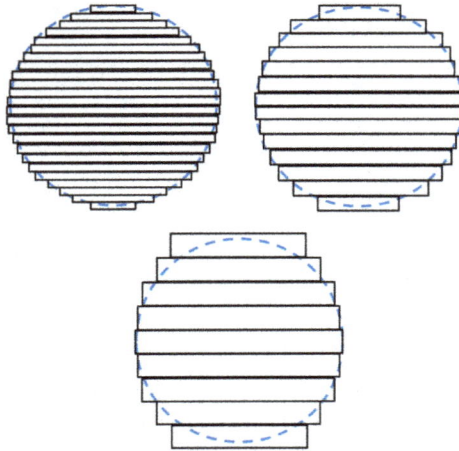

Figure. 4.12 Schematic illustration of variation in the dimensional resolution of a sphere sliced with a different layer thickness.

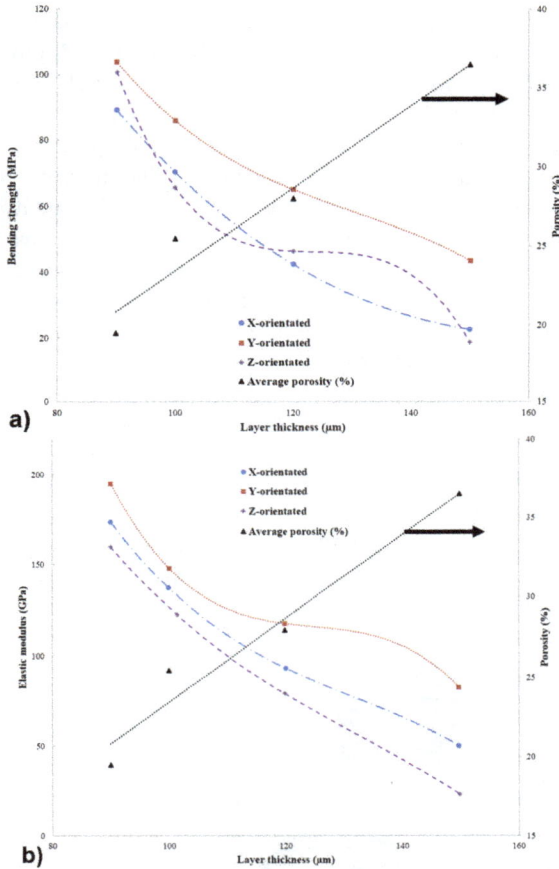

Figure 4.13 Influence of layer thickness on the porosity and mechanical properties of inkjet 3D printed alumina printed in different build ordinations after sintering at 1600 °C (adapted from [30]).

As mentioned in section 4.3.1, in order to obtain as-build parts with desired properties such as green strength, green density, dimensional accuracy, or surface finish, the deposited ink should be spread both laterally and vertically. In addition to the value for SL, layer thickness influences the spread of ink as the required SL is correlated to the volume of powder in each layer. Consider printing mesh structures (see Fig. 2.3) using the liquid BJ on powder bed method [19]. When layer thickness is too thin, the powder volume was insufficient to accommodate the SL utilized. Thus, disproportionate lateral spreading of ink, which stems from ink saturation along the vertical direction, leads to the formation of erratic wires width and thus caused the surface finish to deteriorate. On the other hand, using too large a layer thickness leads to poor ink spreading along the vertical direction decreasing the green strength. At an optimal value for layer thickness, the best dimensional accuracy and surface finish in terms of the wire size and its width variation, together with enough green strength was obtained due to appropriate ink spreading along both lateral and vertical directions. Likewise, enhancing tensile green strength in plaster based feedstock printed under the constant SL was achieved upon decreasing layer thickness at the expense of worsening surface quality [24]. Fig 4.14 presents another example of the trade-off between layer thickness and SL [23]. As can be seen from the figure and for a core-shell saturation ratio of 1 (i.e. A-88, A-100, A-200), increasing layer thickness causes the density of the as-build samples to reduce. Similarly, decreasing the density is the case for the a core-shell saturation ratio of 2 where comparison is made only between the layer thickness of 88 and 200 μm (i.e. B-88, and B-200). Interestingly, for the middle value for layer thickness (100 μm) and at the condition of core-shell saturation ratio of 2, the density of B-100 sample showed the greatest value which arise from sufficient amount of ink along both lateral and vertical directions.

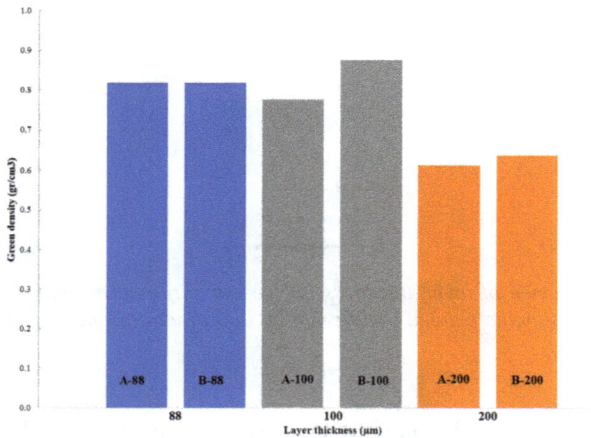

Figure 4.14 Variation in the density of the as-build samples made of calcium sulphate hemihydrate based feedstock with layer thickness and ink saturation level (adapted from [23]).

4.3.3 Drying Time and Power Control

Drying time is the delay time between the ink deposition into a layer and the start of spreading the next layer of powder. Meanwhile, the print head may move to cap cleaning station to assure cleanness of the print head nozzles for reliable ink deposition (see Fig 4.15). Drying the wet layer using an electrical heater is a typical practice in the BJ on powder bed approach which causes initial curing of the binder. The drying time depends on several factors including:

i.   The choice of SL

ii.   Ink type (i.e. binder or solvent), composition of ink

iii. Layer thickness

iv.   Wettability of powder with binder

v.   Powder bed characteristics such as thermal conductivity, surface area, permeability, and packing density

The drying time should be determined before commencing the build cycle. For example, the drying time for metal inkjet 3DP using the liquid BJ on powder bed method could be up to a minute while it is likely to be shorter (e.g. a few second) for the SJ on powder bed approach [31].

Similar to SL and layer thickness, there is an optimal value for drying time [31-33]. Short drying time causes micropores, voids, and cavities to form in the layer due to relative lack of ink-powder interaction [32, 33]. Furthermore, lateral movements of the layer may occur while the subsequent layer is being spread. On the other hand, too long drying time deteriorates intra-layer binding between adjacent layers and therefore distinct layers might form instead of a homogenous part [32, 33]. Additionally, greater drying time extends the build cycle. At optimum period of dry time, ink effectively penetrates through the powder layer vertically and laterally which reduces inter-layer porosity and enhances intra-layer bonding between successive layers which leads to increased green strength and dimensional accuracy of the inkjet 3D printed parts [32, 33].

As mentioned above, partial curing of each layer of powder before spreading a subsequent layer is usually carried out in the liquid BJ on powder bed method using an electrical heater. Drying power control setting or heater power ratio describes the power consumption and temperature of the heater and accordingly the heating rate and the highest temperature of the part being fabricated. It seems that this parameter could play an important part since it controls the drying time of liquid binder and thus control deformation, shrinkage, dimensional accuracy and surface finish of the part during the build cycle. A low power ratio leads to insufficient drying of the liquid binder, causing the build cycle to disrupt or dimensional accuracy and surface finish to deteriorate. High heat dries up the layer very fast and imposes higher deformation and shrinkage within the part, in addition to its adverse impact on the build cycle.

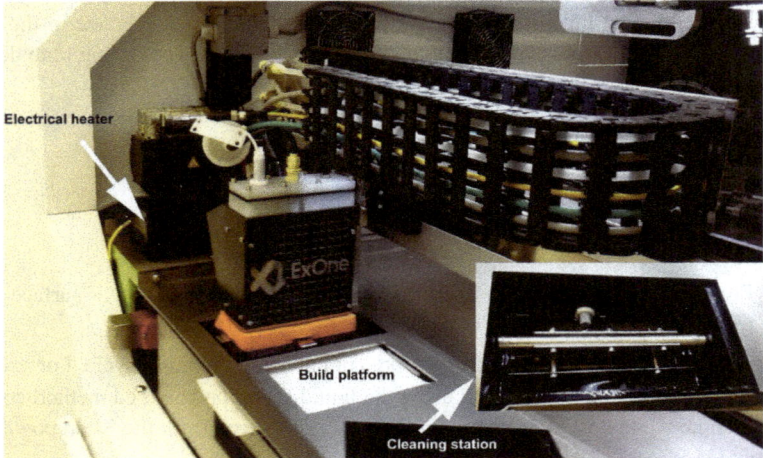

Figure 4.15   The electrical heater and cleaning station within the ExOne Innovent machine used for the liquid BJ on powder bed method.

4.3.4   Build Orientation

The layer stacking orientation refers to the orientation of the part being printed with respect to the powder stacking direction by the roller (i.e. Z-axis) as depicted in Fig. 4.16a. Porosity measurements for amorphous calcium polyphosphate (CPP) powder with various stacking orientations between 0° and 90° indicated that the porosity value along the orientations of 0° and 45° are up to 8% higher than the values for the other orientations (i.e. 30°, 60°, 90°) [34]. In addition, samples with various orientations of the stacked layers showed remarkable differences in mechanical properties, more than three folds [34]. Thus, it seems that this parameter could contribute to porosity percentage as well as mechanical properties of inkjet 3D printed parts. However, the CPP powder used in this study had a particle size between 75 and 150 μm with irregular shape so that the layers of powder with preferred particle orientations forms while these irregular shaped particles being spread by the roller as reported elsewhere [35]. Accordingly, these samples manufactured with preferred particle orientation showed anisotropic mechanical properties [35]. Given that finer particles size with spherical particle shape are utilized for metal inkjet 3DP, it may be less likely to assume that the layer stacking orientation has a substantial effect on the porosity level or mechanical properties in metal inkjet 3DP.

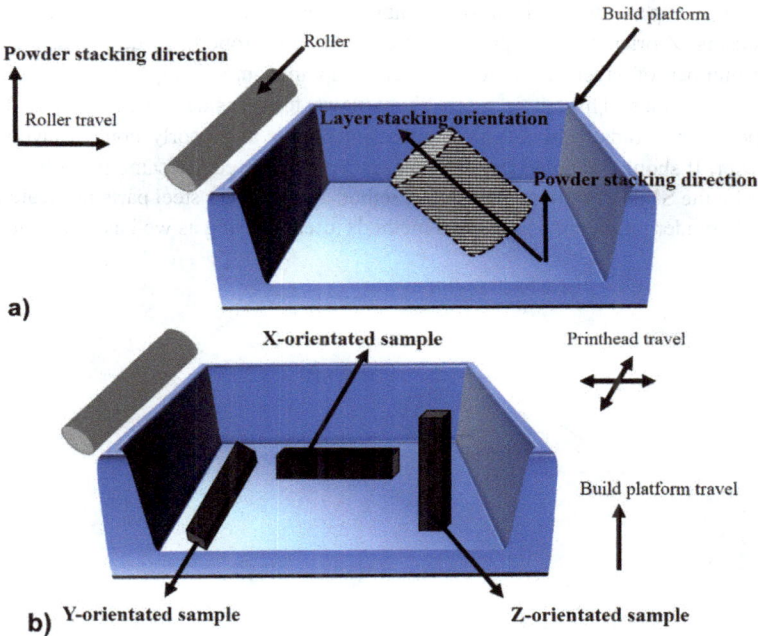

Figure 4.16 Schematic of building chamber and corresponding machine directions: a) layer stacking orientation and b) part build orientation.

The build orientation is the alignment of the parts being printed with respect to the X-, Y- and Z-axis of the build platform as illustrated in Fig. 4.16b whereby X- and Y- axis are the direction of the print head travel and Z-axis is the direction of the build platform move. It has been demonstrated that the build orientations have remarkable impact on the properties of the fabricated parts [30, 33, 36-38]. Anisotropic properties of the printed parts are illustrated in Fig. 4.13. Interestingly, the anisotropic behaviour in terms of fracture toughness was the case even for inkjet 3D printed and sintered Alumina samples followed by infiltration with glass and subsequent sintering as shown in Fig. 4.17. From both Fig. 4.13 and 4.17, Y-orientated samples revealed considerably greater properties and lower porosity percentages due to the deposition of more uniform layer of powder along the Y-axis [30]. This anisotropic behaviour clearly highlights the importance of powder segregation which was addressed in chapter 3. However, further thorough investigation upon this anisotropic behaviour is required to unlock all possible mechanisms. Z-orientated samples indicate the lowest properties due to the fact that a higher number of layers need to be stacked up in comparison with either X or Y-orientated samples. Thus, it is more likely flaws form inside the samples stem from micropores and voids formation throughout inter-layer and poorly bonded layers along intra-layer. It should be noted that the authors have observed the same trend in Ti alloys printed by the SJ on dry mixed feedstock method and stainless steel parts fabricated using the liquid binder jetting on powder bed methods after sintering as well as infiltration.

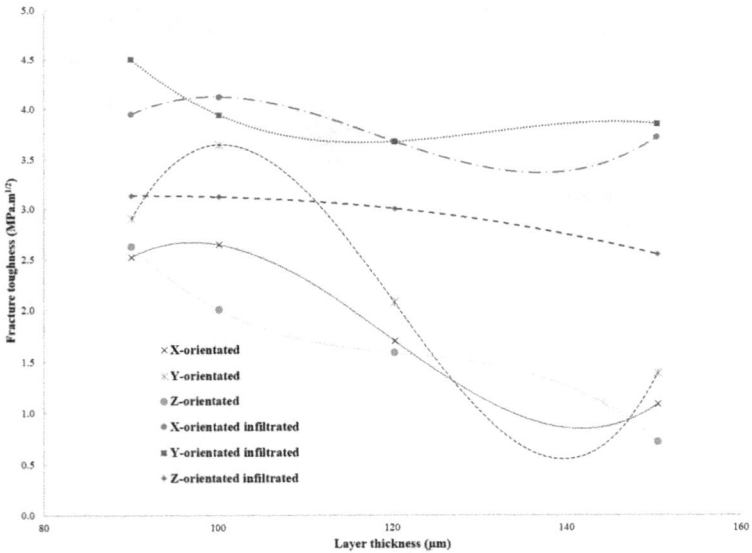

Figure 4.17 Fracture toughness of inkjet 3D printed alumina samples and sintered together with the infiltrated samples with glass followed by subsequent sintering (adapted from [30]).

4.3.5 Roller Spread Speed and Feed-to-Powder Ratio

The translation pace of the roller to spread powder across the powder bed refers to roller spread speed or spread speed. The input ranges of spread speed are from 0.1 to 14 mm/sec, and the suitable speed is currently determined by a trial and error method based on visual cues. Characteristics of powder feedstock such as particle size, particle size distribution, and particle shape determine the spread speed. Higher spread speeds may lead to formation of non-uniform powder bed, especially for the case of powder feedstock with finer particle size. Despite the increasing build time, a lower spread speed (e.g. 1 mm/sec) ensures uniformity across the powder bed.

Feed powder to layer thickness ratio, also known as feed-to-powder ratio, is the ratio of thickness of feed powder to layer thickness. The ratio greater than 2 reassures adequate feedstock supply while spreading a new layer to achieve an evenly spread layer. However, too high a ratio causes layers shifting to occur. As mentioned earlier (section 3.2.2), powder packing density is affected by powder deposition conditions dictated by the roller spread speed and feed-to-powder ratio. Enhancement in sintered flexural strength of liquid BJ on powder bed printed 316L Stainless Steel samples as result of either increasing feed-to-powder ratio from 1 to 3 or decreasing roller spread speed to 6 mm/sec from 10 mm/sec draws attention to the importance of these parameters [39].

4.4 Post-Printing Process

After successful build cycle, the as-built part is a green part embedded in a loose powder within the powder bed (see Fig. 4.18). Post-printing processes are conducted to:

i.   Remove as-built parts from the powder bed

ii. To remove the excess loose powder trapped in the parts

iii. To remove binder materials

iv. To enhance the density and mechanical properties of the parts

More details of each step are provided in the forthcoming subsections.

4.4.1 Post-Printing Bed Manipulation

Before extracting the as-built parts from the powder bed, some binder materials take advantage of a bed manipulation step. As mentioned earlier, curing is required for further polymerization of some binders to strengthen the as-built parts. Other bed manipulations such as drying may also be carried out. It should be noted that any treatment will only influence the regions demarcated by the print head.

Figure 4.18 As-build samples for Charpy impact test made of pure Ti powder which was printed by the SJ on dry mixed feedstock method.

4.4.2 Depowdering

Removing all remnants of the unbound powder from the as-built parts is referred to a depowdering step. After removing the as-built parts from the powder bed, brushing, compressed air blowing, vibration, and vacuuming is conducted depending on the complexities and internal features of the parts (see Fig. 4.19). It is noted that any loose powder embedded after depowdering becomes a part of the final products upon the sintering process.

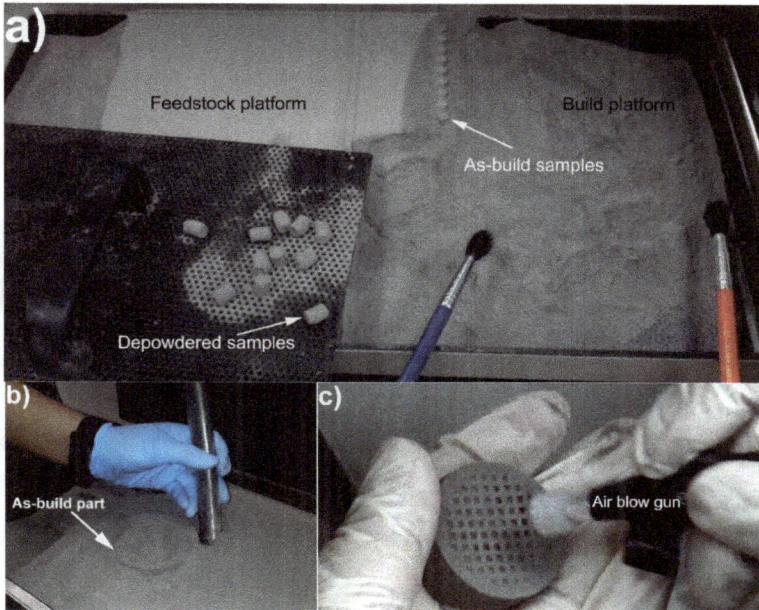

Figure 4.19. Pictures showing: a) removal of as-build samples for compression test made of pure Ti powder by brushing, b) removal of as-build stator made of 420 stainless steel by vacuuming, and c) removal of loose powder from Ti scaffolds using air blowing.

4.4.3 Settering

Repacking as-build parts in refractory powder may be conducted after the depowdering step to prevent the parts from gravitational slumping or other forms of distortion associated with debinding and sintering steps by providing supports especially for unsupported sections. This repacking technique is known as settering and can be seen in Fig 4.20.

4.4.4 Debinding and Sintering

Subsequent thermal treatments consisting of debinding and sintering are optional post-printing steps. However, these steps are mandatory for metal inkjet 3DP. The goal of the debinding step is to progressively burn out the binder materials. A proper debinding profile results from low heating rate combined with single or multiple dwelling time to ensure uniform decomposition and release of the gaseous by-products originated from binder decomposition. Debinding process at higher temperature can be considered as an effective practice if the presented gaseous by-products will not react with the metal powder. Moreover, fast release of the gaseous by-products associated with conducting the process at higher temperatures may lead to distortion, cracking or complete failure of the part. Fig. 4.21 shows part distortion due to an improper debinding cycle.

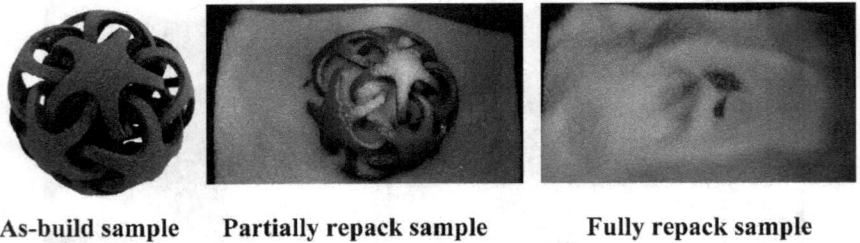

As-build sample   **Partially repack sample**   **Fully repack sample**

Figure 4.20. Settering of stainless steel parts printed by the liquid BJ on powder bed method in the alumina powder with the mesh size of 60.

Figure 4.21 Distortion of Ti based scaffolds due to improper debinding process.

The thermal profile of debinding is specific to each binder. Proper debinding profile can be found using thermal analysis methods. Fig. 4.22 displays a typical example of Thermogravimetric analysis (TGA) for a PVA binder. Such a TGA diagram provides valuable information for the debinding purpose. As can be seen with ramp up of 5 °C /min, the decomposition of PVA reflected by weight loss in the TGA diagram is trivial below 220 °C, while the decomposition rate (i.e. the first derivative of the curve or the slope of the tangent line) for temperatures ranging from 240 °C to 350 °C is maximum and more than 70 % of debinding occurs in this range. Moreover, the TGA curve demonstrates that 97.2 % of PVA binder burns out below 700 °C. According to this result, several debinding profiles can be designed while drawing attention to the following key considerations. The heating rate from room temperature to 220 °C can be as high as 5 °C/min, whereas the heating rate between 240 °C to 350 °C must be slow enough so that gaseous by-products can release uniformly throughout the entire part without adversely affecting its integrity. Practically, slow ramp up combined with single or multiple dwelling times ensure slow and uniform debinding throughout the parts. In order to fine-tune the debinding profile, several designed profiles can be examined using TGA analysis.

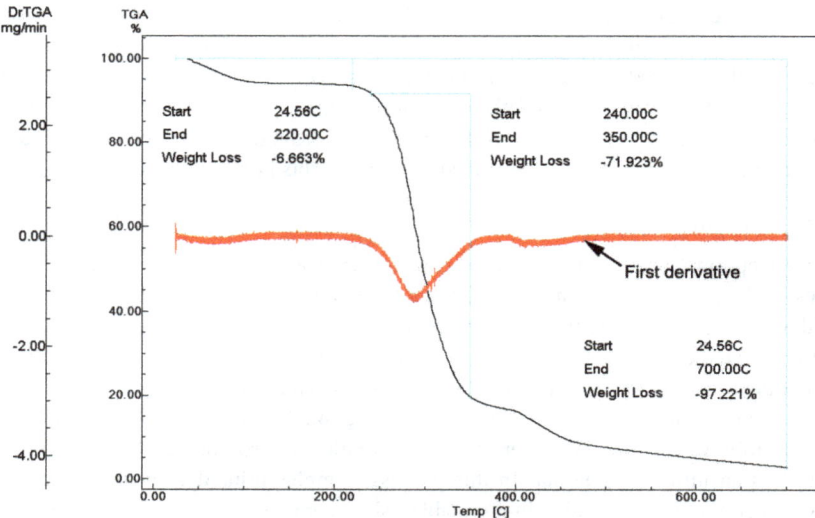

Figure 4.22 Thermogravimetric analysis (TGA) of polyvinyl alcohol (PVA) binder under an inert atmosphere of argon with ramp rate of 5 °C/min.

Essentially, the key issues related to the debinding step include:

i.   It is a lengthy and messy process.

ii. It causes the as-build parts to shrink leading to dimensional changes.

iii. A residue of binder may remain in the part which might be considered as contamination for parts' functionality.

iv. Parts may collapse or distort after the debinding step and before subsequent sintering.

The debinding process is followed by sintering in which binder free constructs having low initial density densify to higher final density. Metals have tremendous sintering flexibility so that it is possible to tailor sintering behaviour to suit the parts' requirements. In this regard, well established knowledge of the conventional powder metallurgy (PM) process should be taken into account as the main guide line for successful sintering. However, it is plausible that inkjet 3D printed parts exhibit completely different sintering behaviour when compared to anticipated sintering behaviour in conventional PM fabricated samples. This inconsistency arises from the absence of compaction in inkjet 3D printed parts when compared to traditional PM parts. For example, in the case of PM Al alloys, mechanical force is usually applied to compact samples. Thus, stable oxide film on the surface of Al particles, which hinders diffusion process during sintering, is mechanically ruptured by plastic deformation of Al particles during the compaction step. As the traditional powder compaction step during inkjet 3DP is missing, it is natural to expect challenges to sinter 3D printed parts. Accordingly, other techniques can be considered for sintering such as described in some patents [40, 41].

4.4.5 Infiltration

Another optional post-printing process is to infiltrate a metal or polymer into the open pores of the printed parts. The infiltrant materials must have a lower melting point than solidus temperature or melting temperature of the parts. In addition, the parts should be partially sintered and strengthened for metal infiltration. The main purpose of infiltration is to obtain higher density parts with superior mechanical properties and minimize dimensional changes associated with sintering. Fig. 4.23a shows a single step debinding process followed by partial sintering and infiltration of iron-based printed parts with bronze as an infiltrant material. In this process, samples printed with the liquid BJ on powder bed method are loaded in a crucible using alumina as supporting powder for the infiltration process. The binder was first removed from the printed parts using furnace heating which was followed by partial sintering of the parts. The temperature was further increased to start infiltration of the parts by melting the bronze powder. Transportation of the molten bronze to the printed samples was realized through the runners and gates. Finally, the infiltrated samples were removed from the crucible as can be seen in Fig. 4.23b.

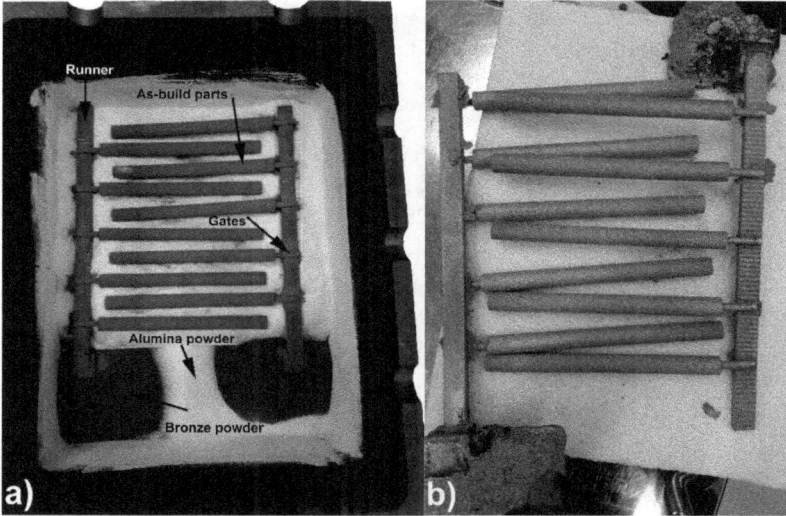

Figure 4.23 a) Single step debinding and infiltration of liquid binder jet printed 420 stainless steel samples with bronze conducted by ExOne company and b) 420 stainless steel infiltrated with bronze.

Summary

This chapter provides an overview of the inkjet technology along with the determination of printing and post-printing related parameters. Ink droplets with a variety of size can be generated by the print heads operating on continuous inkjet and drop-on-demand (DOD) mechanisms. Thermal and piezoelectric DOD inkjet print heads are the two most common methods of choice for projecting ink into the powder bed. When ink drops hit the powder bed, they deform and intermingle with particles. These steps followed by the absorption step result in the formation of particle aggregates having dimensions of several times of either powder particle size or droplet size. In essence, ink saturation level, layer thickness, and drying settings contribute in controlling lateral and vertical interaction between ink and the powder bed, thereby deciding properties of as-build and sintered parts. The roller spread speed and feed-to-powder ratio determine the uniformity of powder spreading along the powder bed. However, obtained properties of printed parts are dependent upon the build orientation, regardless of their conditions (i.e. as-build, sintered and infiltrated). Post-printing processes commence with bed manipulation followed by depowdering and optional settering steps. Debinding and sintering steps transform the as-build parts into end parts whereby binder materials are burnt out and the parts densify. The infiltration step makes it possible to achieve further densification.

References

1. Martin, G.D. and I.M. Hutchings, Fundamentals of Inkjet Technology, in Inkjet Technology for Digital Fabrication. 2012, John Wiley & Sons, Ltd. p. 21-44. https://doi.org/10.1002/9781118452943.ch2

2. Heinzl, J. and C.H. Hertz, Ink-Jet Printing, in Advances in Electronics and Electron Physics, W.H. Peter, Editor. 1985, Academic Press. p. 91-171. https://doi.org/10.1016/S0065-2539(08)60877-X

3. Derby, B., Inkjet printing of functional and structural materials: Fluid property requirements, feature stability, and resolution, in Annual Review of Materials Research. 2010. p. 395-414. https://doi.org/10.1146/annurev-matsci-070909-104502

4. Samuel, J. and P. Edwards, Solvent-Based Inkjet Inks, in The Chemistry of Inkjet Inks. 2011, WORLD SCIENTIFIC. p. 141-159.

5. Le, H.P., Progress and trends in ink-jet printing technology. Journal of Imaging Science and Technology, 1998. 42(1): p. 49-62.

6. Ruiz, O.E. CFD model of the thermal inkjet droplet ejection process. in 2007 Proceedings of the ASME/JSME Thermal Engineering Summer Heat Transfer Conference - HT 2007. 2007.

7. Calvert, P. and T. Boland, Biopolymers and Cells, in Inkjet Technology for Digital Fabrication. 2012, John Wiley & Sons, Ltd. p. 275-305. https://doi.org/10.1002/9781118452943.ch12

8. Yeates, S.G., et al., Fluids for Inkjet Printing, in Inkjet Technology for Digital Fabrication. 2012, John Wiley & Sons, Ltd. p. 87-112. https://doi.org/10.1002/9781118452943.ch4

9. Murphy, S.V. and A. Atala, 3D bioprinting of tissues and organs. Nat Biotech, 2014. 32(8): p. 773-785. https://doi.org/10.1038/nbt.2958

10. Cui, X., et al., Cell damage evaluation of thermal inkjet printed Chinese hamster ovary cells. Biotechnology and Bioengineering, 2010. 106(6): p. 963-969. https://doi.org/10.1002/bit.22762

11. Akira, A., H. Toshitami, and E. Ichiro, One-Dimensional Model of Bubble Growth and Liquid Flow in Bubble Jet Printers. Japanese Journal of Applied Physics, 1987. 26(10R): p. 1794.

12. Chen, P.-H., W.-C. Chen, and S.H. Chang, Bubble growth and ink ejection process of a thermal ink jet printhead. International Journal of Mechanical Sciences, 1997. 39(6): p. 683-695. https://doi.org/10.1016/S0020-7403(96)00067-7

13. Dijksman, J.F., et al., Precision ink jet printing of polymer light emitting displays. Journal of Materials Chemistry, 2007. 17(6): p. 511-522. https://doi.org/10.1039/B609204G

14. Tekin, E., P.J. Smith, and U.S. Schubert, Inkjet printing as a deposition and patterning tool for polymers and inorganic particles. Soft Matter, 2008. 4(4): p. 703-713. https://doi.org/10.1039/b711984d

15. Stringer, J. and B. Derby, When the Drop Hits the Substrate, in Inkjet Technology for Digital Fabrication. 2012, John Wiley & Sons, Ltd. p. 113-139. https://doi.org/10.1002/9781118452943.ch5

16. Fan, T., Droplet-powder impact interaction in three-dimensional printing. 1995 Massachusetts Institute of Technology: Cambridge.

17. Lanzetta, M. and E. Sachs, Improved surface finish in 3D printing using bimodal powder distribution. Rapid Prototyping Journal, 2003. 9(3): p. 157-166. https://doi.org/10.1108/13552540310477463

18. Bai, Y. and C.B. Williams, An exploration of binder jetting of copper. Rapid Prototyping Journal, 2015. 21(2): p. 177-185. https://doi.org/10.1108/RPJ-12-2014-0180

19. Lu, K. and W.T. Reynolds, 3DP process for fine mesh structure printing. Powder Technology, 2008. 187(1): p. 11-18. https://doi.org/10.1016/j.powtec.2007.12.017

20. Oriakhi, C. and T.M. Lambright, Calcium aluminate cement compositions for solid freeform fabrication. 2007, Google Patents.

21. Monkhouse, D.C., et al., Rapid prototyping and manufacturing process. 2003, Google Patents.

22. Castilho, M., et al., The role of shell/core saturation level on the accuracy and mechanical characteristics of porous calcium phosphate models produced by 3Dprinting. Rapid Prototyping Journal, 2015. 21(1): p. 43-55. https://doi.org/10.1108/RPJ-02-2013-0015

23. Suwanprateeb, J., et al., Influence of printing parameters on the transformation efficiency of 3D-printed plaster of paris to hydroxyapatite and its properties. Rapid Prototyping Journal, 2012. 18(6): p. 490-499. https://doi.org/10.1108/13552541211272036

24. Vaezi, M. and C.K. Chua, Effects of layer thickness and binder saturation level parameters on 3D printing process. The International Journal of Advanced Manufacturing Technology, 2011. 53(1): p. 275-284. https://doi.org/10.1007/s00170-010-2821-1

25. Szucs, T.D. and D. Brabazon, Effect of saturation and post processing on 3D printed calcium phosphate scaffolds, in Key Engineering Materials. 2009. p. 663-666. https://doi.org/10.4028/www.scientific.net/KEM.396-398.663

26. Nandwana, P., et al., Powder bed binder jet 3D printing of Inconel 718: Densification, microstructural evolution and challenges. Current Opinion in Solid State and Materials Science, 2017. https://doi.org/10.1016/j.cossms.2016.12.002

27. Sachs, E.M., et al., Metal and ceramic containing parts produced from powder using binders derived from salt. 2003.

28. Utela, B.R., et al., Development Process for Custom Three-Dimensional Printing (3DP) Material Systems. Journal of Manufacturing Science and Engineering, 2010. 132(1): p. 011008-011008. https://doi.org/10.1115/1.4000713

29. Shanjani, Y., E. Toyserkani, and C. Wei. Modeling and characterization of biomaterials spreading properties in powder-based rapid prototyping techniques. in ASME International Mechanical Engineering Congress and Exposition, Proceedings. 2008.

30. Zhang, W., et al., Three-dimensional printing of complex-shaped alumina/ glass composites. Advanced Engineering Materials, 2009. 11(12): p. 1039-1043. https://doi.org/10.1002/adem.200900213

31. Chen, H. and Y.F. Zhao, Process parameters optimization for improving surface quality and manufacturing accuracy of binder jetting additive manufacturing process. Rapid Prototyping Journal, 2016. 22(3): p. 527-538. https://doi.org/10.1108/RPJ-11-2014-0149

32. Farzadi, A., et al., Effect of layer printing delay on mechanical properties and dimensional accuracy of 3D printed porous prototypes in bone tissue engineering. Ceramics International, 2015. 41(7): p. 8320-8330. https://doi.org/10.1016/j.ceramint.2015.03.004

33. Asadi-Eydivand, M., et al., Effect of technical parameters on porous structure and strength of 3D printed calcium sulfate prototypes. Robotics and Computer-Integrated Manufacturing, 2016. 37: p. 57-67. https://doi.org/10.1016/j.rcim.2015.06.005

34. Vlasea, M., R. Pilliar, and E. Toyserkani, Control of structural and mechanical properties in bioceramic bone substitutes via additive manufacturing layer stacking orientation. Additive Manufacturing, 2015. 6: p. 30-38. https://doi.org/10.1016/j.addma.2015.03.001

35. Shanjani, Y., et al., Mechanical characteristics of solid-freeform-fabricated porous calcium polyphosphate structures with oriented stacked layers. Acta Biomaterialia, 2011. 7(4): p. 1788-1796. https://doi.org/10.1016/j.actbio.2010.12.017

36. Castilho, M., et al., Fabrication of computationally designed scaffolds by low temperature 3D printing. Biofabrication, 2013. 5(3). https://doi.org/10.1088/1758-5082/5/3/035012

37. Yao, A.W.L. and Y.C. Tseng, A robust process optimization for a powder type rapid prototyper. Rapid Prototyping Journal, 2002. 8(3): p. 180-189. https://doi.org/10.1108/13552540210431004

38. Farzadi, A., et al., Effect of Layer Thickness and Printing Orientation on Mechanical Properties and Dimensional Accuracy of 3D Printed Porous Samples for Bone Tissue Engineering. PLOS ONE, 2014. 9(9): p. e108252. https://doi.org/10.1371/journal.pone.0108252

39. Shrestha, S. and G. Manogharan, Optimization of Binder Jetting Using Taguchi Method. JOM, 2017. 69(3): p. 491-497. https://doi.org/10.1007/s11837-016-2231-4

40. Foerster, G.S. and G.D. Lawrence, Sintering of loose particulate aluminum metal. 1967.

41. Liu, J., Processes for sintering aluminum and aluminum alloy components. 2009.

CHAPTER 5

Overview of Inkjet 3D Printed Metals and Their Key Performance Indicators

Abstract

This chapter focuses on metallic systems developed using inkjet 3DP and their end properties. It presents the overview of monolithic, composite, and functionally graded metallic materials together with their printing and post-printing related parameters. Density, shrinkage, resolution, surface roughness, porosity related feature, biological, and mechanical properties of these metallic materials along with influential parameters for each property are evaluated as key performance indicators for the inkjet 3DP technology for metals. The addressed indicators can serve as a good assessment for design related parameters which enable the reader to judiciously regulate them in order to achieve the intended properties.

Keywords

Inkjet 3DP, Metals and Alloys, Density, Shrinkage, Pores, Resolution, Surface Roughness, Biological Properties, Mechanical Properties, Printing Parameters, Sintering

5.1 Overview of Developed Metallic Systems

5.1.1 Monolithic Metals and Alloys

Inkjet 3DP has been applied to different metals and alloys in order to fabricate components for the injection molding industry, thermal management systems, structural electronics, biomedical applications etc. Inkjet 3D printed metals and alloys, along with their printing details and sintering profiles are tabulated in Table 5.1. As can be seen from the table, material systems mainly include Ti alloys, stainless steels, and nickel based superalloys, which have been printed using the SJ on dry mixed feedstock and the liquid BJ on powder bed methods. The layer thickness of 100 μm and particle size smaller than 45 μm can be considered as the trend in these systems. For the sintering profile, however, a broad variety of profiles in terms of temperature and time have been employed for the same material systems.

Table 5.1 Material systems used in metal inkjet 3DP.

Powder system	Method	Printing process		Binder materials				Sintering profile	Ref
		Layer thickness (μm)	Metal powder size (μm)	Name	Content	Size (μm)	Solvent		
Pure Ti	SJ on dry mixed feedstock	100, 125,175	45-150, 106-150, 75-106, <75 and >45	PVA	5 %	<63	Zb^{TM}58	1100 °C and 1400°C for 1-3 hours	[1]
Pure Ti	SJ on dry mixed feedstock	-	<45	PVA	-	75	-	900-1350 °C	[2]
Pure Ti	SJ on dry mixed feedstock	100	45	PVA	5 and 10 wt.%	-	-	1250-1370 °C for 2 and 5 hours	[3]
Pure Ti	SJ on dry mixed feedstock	100	<45	PVA	10 wt.%	-	-	1000-1350 °C for 2 hours	[4]
Pure Ti	SJ on dry mixed feedstock	-	-	PVA	-	75	-	1250-1350 °C for 2 hours	[5]
Pure Ti	SJ on dry mixed feedstock	-	<45	PVA	-	75	-	900-1300 °C	[6]
Pure Ti	SJ on dry mixed feedstock	100	<74	PVA	10 %	125	-	1200-1400 °C	[7]
Pure Ti	SJ on dry mixed feedstock		45-75	PVA	3 %	-	Zb^{TM}58	1400 °C	[8]
Pure Ti	SJ on dry mixed feedstock	-	45	PVA	5 and 10 %	-	-	1250-1370 °C for 2-5 hours	[9]
Ti-5Ag	Salt solutions jetting on powder bed	-	60	Aqueous solution of AgNO$_3$	5.8 Molar	-	-	1150 and 1300 °C for 1 hour	[10]
17-4PH stainless steel	SJ on granulated feedstock	50-200	D$_{90}$< 16μm	-	-	-	-	1330 °C for 3 hours	[11]
316L stainless steel	Liquid BJ on powder bed	75-100	13, 17, 38, 73	PM-B-SR2-02	-	-	-	1200-1405 °C for 1.5 hours	[12]
420 stainless steel	Liquid BJ on powder bed	-	30	-	-	-	-	1100-1300 °C for 1.5-20 hours	[13]
420L stainless steel	Liquid BJ on powder bed	100	30	-	-	-	-	1120 °C	[11]

Powder system	Method	Layer thickness (μm)	Metal powder size (μm)	Printing process				Sintering profile	Ref
				Binder materials					
				Name	Content	Size (μm)	Solvent		
Inconel 718	Liquid BJ on powder bed	100-200	16.5 and 26	-	-	-	-	1260-1300 °C for 4 hours	[14]
Inconel 718	Liquid BJ on powder bed	125	6.1-53	WS-24E and PM-B-SRX	-	-	-	1290-1300 °C for 2 hours	[15]
Inconel 718	Liquid BJ on powder bed	>75	$D_{50} = 7$, 20, and 66	A proprietary aqueous binder, based on diethyl glycol	70 % and 80 %	-	-	1185-1330 °C for 5 hours	[16]
Inconel 625	Liquid BJ on powder bed	100	16-53	Commercial binder consisting of ethylene glycol monobutyl ether (10 vol.%) and ethylene glycol (20 vol.%)	-	-	-	1200-1300 °C for 4 hours	[17-19]
316L stainless steel	Liquid BJ on powder bed	-	60	Acrysol WS-2	-	-	-	-	[20]
316L stainless steel	Liquid BJ on powder bed	75-150	20 and 75	Standard water-based binder	-	-	-	1280 °C for 2 hours	[21]
316L stainless steel	Liquid BJ on powder bed	50 and 100	$D_{90} < 22$	ExOne M-Flex binder	-	-	-	1360 °C for an hour	[22]
Ti-50.5 at% Ni	SJ on granulated feedstock	50-200	Granule size < 100	-	-	-	-	1200 °C for 2 hours	[23]
Fe-30 Mn	Liquid BJ on powder bed	100	<44	-	-	-	-	1200 °C for 3 hours	[24]
Fe-30 Mn-1 Ca	Liquid BJ on powder bed	-	<44	A water-based organic binder	-	-	-	1200 °C for 3 hours	[25]
Cu	Liquid BJ on powder bed	80 and 100	15 and 75	-	-	-	-	1080-1090 °C for 2-4 hours	[26]
Carbonyl nickel	SJ on coated feedstock and BJ on powder bed	100 and 200	5	Starch and PVA	-	-	ZB4	600-1350 °C for 1-10 hours	[27]
Ti35Ni50Hf15	Liquid BJ on powder bed	20-50	5.5	An acrylic-based aqueous binder	-	-	-	-	[28]
Ti35Ni50Hf15	Liquid BJ on powder bed	40-300	20-150	An acrylic-based aqueous binder	-	-	-	-	[29]
Ag	BJ on powder bed	25	1.2-2	Water-based organic	-	-	-	850 °C for 20 min	[30]

Powder system	Printing process								Sintering profile	Ref
	Method	Layer thickness (μm)	Metal powder size (μm)	Binder materials						
				Name	Content	Size (μm)	Solvent			
Co-28Cr-6Mo	Liquid BJ on powder bed	75-150	20 and 75	binder + 20 % silver nanoparticle Standard water-based binder	-	-	-		1280 °C for 2 hours	[21]
Co-28Cr-6Mo	Liquid BJ on powder bed	130	20 and 64	PM-B-SR2-02	-	-	-		1280 °C for 2 hours	[31]

5.1.2 Composites and Functionally Graded Materials

Processing a metal matrix composite system can easily be conducted through inkjet 3DP. For example, 420 stainless steel powder with the mean size of 35 μm was dry mixed with different contents of 2 μm sized Si_3N_4 powder as a reinforcement. This feedstock was then fed into the feed platform, and similar printing procedure and post-printing processes were applied to obtain the metal matrix composites [32, 33]. Likewise, inkjet 3DP was successfully employed to fabricate Ti-hydroxyapatite (HA) functionally graded materials comprising two segments of 100% Ti and 80% Ti-20% HA [34]. The starting materials were nano-sized Ti and HA particles with average diameters of 50 nm and 40 nm, respectively. These materials were blended together with PVA in the ratio of Ti: HA: PVA: 8:2:1 (mass) to infuse into the feed platform. Microstructural results showed that despite the sharp interface between the printed segments, no crack emerged after sintering at 1200 °C. However, there is no data on the evaluation of reproducibility of such a composite material because segregation would be a real issue for this kind of feedstock as highlighted under the segregation section (3.2.4) in chapter 3.

Three-dimensional variations in material compositions on the surface or inside of constructs with a complex shape can be fabricated in a single process via adding consecutive layers of desired materials. With the aid of upgrading the hardware to jet different functional ink through different print heads together with software modifications to prescribe locations of each print head, graded structures of carbon steel were able to be printed [35]. In view of this, various suspensions of ink consisting of nano-sized carbon black particles and acrylic binders were dispensed through multiple print heads into the X30Cr13 powder bed layer by layer to fabricate the graded as-built parts in which the concentrations of carbon black was varied along the deposited layers. After burning the binder and sintering, the carbon element of the ink remained in the ferrous matrix, locally acting as an alloying element which resulted in graded variations of carbon between 0.35% and 0.83% in the end parts. It should be pointed out that several lab-scaled systems based on the inkjet technology have been developed which can be used to introduce functionally graded structure in metallic materials. As an example, a developed mechatronic platform called a Variable Density inkjet 3DP system allows three different types of powder feedstock to be deposited with in the powder bed along with control over localized binder deposition [36, 37].

5.2 Key Performance Indicators of Inkjet 3D Printed Metal Parts

Density, shrinkage and dimensional accuracy, porosity related properties, biological properties, and mechanical properties of inkjet 3D printed metallic parts along with influential parameters on each property are described in the forthcoming sections. Theses sections addresses the strengths and weaknesses of the inkjet 3DP technique for AM of metal based systems.

5.2.1 Density

Density of the as-built parts is between bulk and tap density of raw powder as mentioned in chapter 3. However, one researcher reported that the relative density of the as-built samples can be higher than the tap density of powder (i.e. approximately 10% higher) which could be attributed to the printing parameters used in this study such as the thin layer thickness of 50 μm, printing method, binder curing process etc. [22]. Nevertheless, all researchers subscribe to the view that the relative density of the as-built parts is less than 60%, usually about 50%. Cross sectional micro-computed tomography (microCT) image of as-built IN 625 samples printed using the liquid BJ on powder bed method with the layer thickness of 100 μm is shown in Fig. 5.1 [17]. The overall relative density of this sample was approximately 53-60%. The parallel lines indicate each deposited layer of the feedstock in which a higher porosity level is discernible. These parallel lines can be modified by controlling printing parameters to achieve better intra-layer binding as displayed in Fig 3.3.

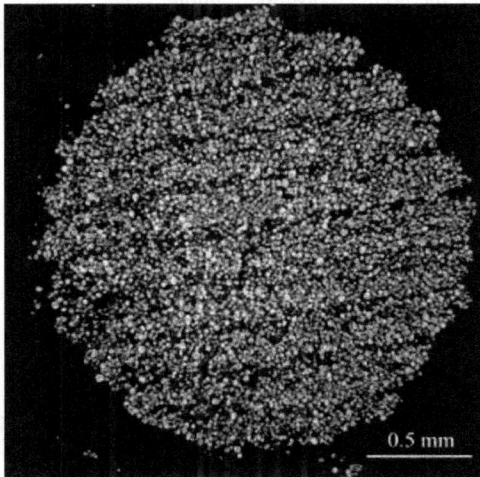

Figure 5.1 Cross sectional microCT image of as-built IN 625 samples printed using liquid BJ on powder bed methods with the layer thickness of 100 μm (courtesy – A.Mostafaei [17]).

Printing methods directly affect the green density and thus the final density of the parts. Among the aforementioned printing methods, the SJ on powder bed approach provides lower green density because the binder materials within the powder feedstock act as a space holder. Thus, these methods can be used to fabricate porous structures when pores serve as an advantage [7, 8, 38]. The final relative density ranging from 55 to 99.8% can be accomplished in printed parts using the BJ on powder bed approach [12, 14, 17, 18, 22, 26, 39]. For instance, Fig. 5.2a shows microstructure of IN 718 printed with the liquid BJ on powder bed method after sintering at 1280 °C, indicating fine-grained structure as well as the relative density of approximately 98% [14]. As another example, Fig. 5.2b displays liquid BJ on powder bed printed Inconel 625 samples with relative density of 99.6 % after sintering at 1280 °C for four hours while the relative density of the as-built samples prior to sintering was 53% [19]. For the SJ on granulated feedstock, the relative density of 95% was obtained for printed parts made of 17-4PH stainless steel and NiTi after sintering at 1330 °C and 1200°C, respectively [11, 23]. However, it is by no means certain which of these printing methods provide higher density. There is insufficient data on the SJ on coated feedstock method to conclude the capability of this method. On the one hand, the starch coating contributes toward strong cohesion and bonding between the metal powder, resulting in good sinterability. On the other hand, samples printed with the SJ on coated feedstock method demonstrated inhomogeneous microstructure and density distribution after sintering [27, 40]. Only one value for relative density of this method is available in the literature, which was greater compared to the liquid BJ on powder bed method for the same material system [27].

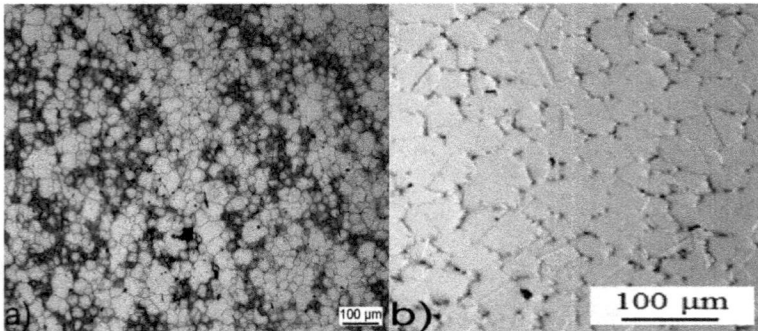

Figure 5.2 The optical microstructure of liquid BJ on powder bed printed nickel-based superalloys after sintering: a) Inconel 718 [14] and b) Inconel 625 (courtesy – A. Mostafaei [19]).

As mentioned in chapter 4, layer thickness has remarkable impact on the relative density of the printed parts and thus their mechanical properties. However, variations of the relative density for IN 718 samples printed with five different layer thicknesses ranging from 100 to 200 μm were about 4% after sintering at 1260 °C [14]. Interestingly, the relative density became virtually independent from the layer thickness with increasing sintering temperature to 1280 °C and 1300 °C [14]. It seems that more comprehensive investigations are required to study the effect of layer thickness on metal inkjet 3DP since it is plausible to assume more significant contribution for layer thickness in relative density of the as-build parts and therefore the sintered parts. Neither the inkjet 3DP apparatus nor the heating rate during sintering are significant factors on the final density of the printed parts and they have only a minor impact on the final density. For example, inkjet 3DP apparatus and sintering heating rate have negligible effects (less than 1.5%) on the sintered density of Inconel 718 samples [15].

Specifications of metal powder and sintering profiles are the other major contributing factors which affect the final density of the inkjet 3D printed constructs. There is a close connection between achievable density and particle size; higher density can be achieved for finer particles at the same sintering temperature due to higher sinterability of fine particles [31]. This was illustrated by Verlee et al. [12] investigations in which the control of density and porosity of printed 316L stainless steel parts was reported. Different mean particle sizes of 13, 38, and 73 μm were used to print the 316L stainless steel parts with approximately the same green density. Interestingly, after sintering at 1405 °C, the relative density ranged from 97.5% to 65.3%. It is worthy to note that inkjet 3DP of fine powder was found to be difficult due to powder spreading and agglomeration issues, which may lead to lower obtained green density than compared to using coarser powder [14]. By using a combination of fine and coarse powders, it is likely this printing issues will diminish and thus enhance densification during sintering [15]. A comprehensive discussion on correlation between characteristics of powder feedstock and its sinterability was made in section 3.2.3.

Sintering is a thermally activated diffusion process in which individual particles are bonded together and contacts between them grow as a result of pores elimination. According to Coble [41], three stages of sintering are:

1) The initial stage: The interparticle contact area increases by neck growth up to 0.2 of the cross-sectional area of particles which results in relative density in the range of 50-60%.

2) The intermediate stage: Continuous pore channels forms at triple points and the relative density increases up to 90%.

3) <u>The final stage</u>: This stage starts when the pore phase is eventually pinched off and the pores shrink continuously to nearly a zero size.

Surface diffusion plays a crucial role during the early stage of sintering: Atoms move to the contacting points through a solid state diffusion process, resulting in necks formation. This is then continued by the necks growth and bands formation between particles as the temperature increases. In this stage, bulk diffusion and grain boundary diffusion take place which make pores shrink and transform into a tunnel-shape linking to other pores. Some small pores may also disappear. Thus, even though some pores might join to form a larger pore, decline in the number of pores and the overall porosity percentage are the dominated phenomena. Pore coarsening and its impact on densification during the final stage of sintering is a well-known phenomenon.

Sintering is a multi-faceted process governed by numerous factors, including powder chemical composition, physical specifications of powder, surface chemistry of particles, consolidation method, green density of compact, sintering profile, and sintering atmosphere. Increasing sintering temperature generally increases density as would be anticipated. Fig. 5.3 represents densification characteristics of inkjet 3D printed gas atomized Inconel 625 samples along with sintering temperature. As can be seen in Fig. 5.4, the relative density of the samples at 1225 °C was 85% and it constantly enhanced with increasing temperature, reaching to more than 99 % in the range of 1285 and 1300 °C [42]. Sintering time is also essential for higher densification. Density increases as the holding time increases [12, 13].

Mostafaei et al. [18] investigated microstructural evaluation of water and gas atomized Inconel 625 powder. As the result of different powder chemistry originated from atomization process, inkjet 3D printed parts demonstrated distinctive relative densities within the range of sintering temperatures. The water atomized powder showed higher relative density in some temperature ranges while the gas atomized one demonstrated greater relative density in the other temperature ranges as shown in Fig. 5.4. This arises from the fact that variation in powder surface chemistry is synonymous with different powder sinterability.

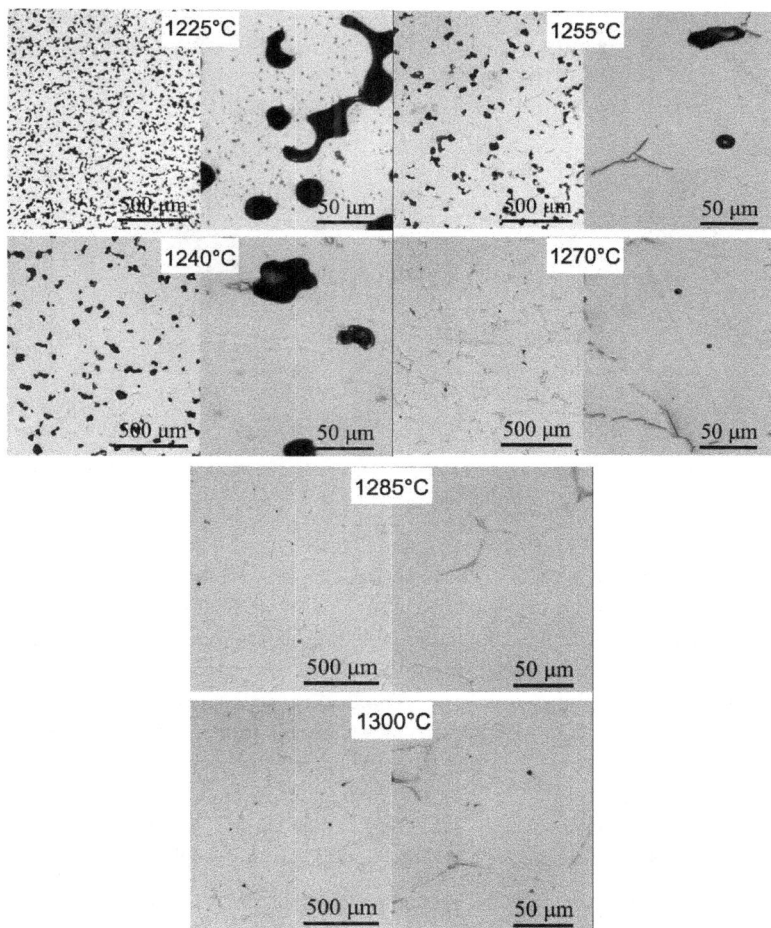

Figure 5.3   The porosity structures of inkjet 3D printed gas atomized Inconel 625 samples with variation in sintering temperature from 1225 °C to 1300 °C [42].

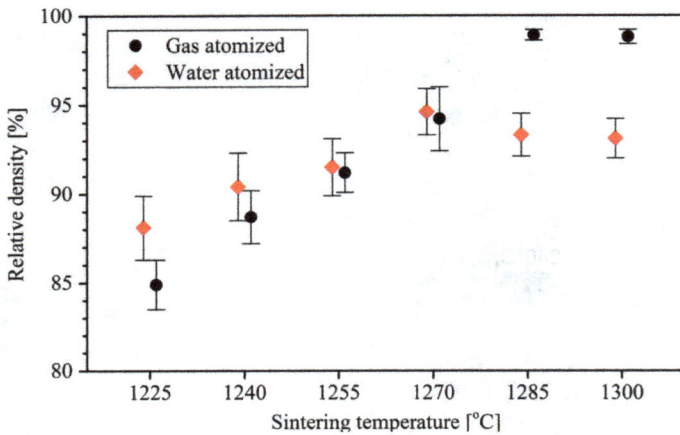

Fig 5.4 Relative density of inkjet 3D printed gas atomized and water atomized Inconel 625 samples plotted against the sintering temperature [42].

The sintering atmosphere can also have a considerable influence on the final relative density. A reducing atmosphere can enhance the purity of final products and affects the sintered density. For example, a hydrogen sintering atmosphere not only enhanced the purity of inkjet 3D printed copper parts up to 4%, but also improved the final density by more than 25% in comparison with non-reducing atmospheres [26]. Furthermore, sintering under vacuum condition may do more benefits to densification compared with gas mixture condition due to alteration in sintering's driving force of pores [13]. Zhou et al. [13] investigated the effect of the sintering atmosphere for inkjet 3D printed 420 stainless steel. The results revealed that the samples sintered under vacuum condition exhibited about 4% higher density than those sintered under gas mixture conditions.

Several attempts have been made to simulate and predict the sintering behavior of powder. A couple of approaches, including continuum theory [43], Monte-Carlo methods [44], and a master sintering curve [45] have been applied to develop the relationship between sintering profiles and final density. Nonetheless, more research needs to be conducted to apply these approaches to the inkjet 3DP domain due to the fact that inkjet 3DP differs in many aspects from other powder processing techniques. For example, the topology of porous structures in terms of pore size distribution and pore shape (morphology) is clearly different for inkjet 3DP and the PM prepared samples. It substantiated that the topological features indubitably influence the sintering kinetics

[46]. Another clear difference is that almost no compaction is utilized in inkjet 3DP and accordingly sintering behavior of loose metal powder is quite different from compacted samples.

It has been found that the addition of some elements can alter the sintering mechanism from solid atomic transport to transient liquid phase sintering at the same sintering temperature. This results in enhanced atomic transport/materials diffusion and thus accelerates the densification process [47, 48]. Sun et al. [49] employed inkjet 3DP to fabricate a highly dense 420 stainless steel part through the addition of Si_3N_4 with the mean size of 2 μm. According to the Si–Fe phase diagram, liquid phases can be produced above 1225°C if the Si content is higher than 6.9 wt%. The higher the silicon content, the lower is the liquidation-beginning temperature, and the higher becomes the densification rate. For instance, at 1298°C, the maximum densification rate for the sample without Si_3N_4 was 21.76 μm/min, whereas that of the sample with 5 wt% Si_3N_4 was 61.31 μm/min [32]. Hence, the relative density continually rose with increasing Si_3N_4 contents up to 12.5 wt % and reached the highest value of 99.8% for the sample sintered at 1300°C. Moreover, the shape of the inkjet 3D printed parts during the sintering process may be stabilized by the presence of Si_3N_4 particles in the feedstock, enhancing the ultimate dimensional accuracy and higher sharp edges quality of the parts [33].

As mentioned above, relative density of the as-built parts is about 60%. Sintering to full density of large as-built parts with such a low relative density is a problematic issue due to problems with maintaining dimensional accuracy. Moreover, in order to increase the parts' strength and to remove residual porosity, several post-printing processes may be required. Hot isostatic pressing (HIP) is one of the possible techniques to reach full density in inkjet 3D printed parts. Artalejo et al. [15] performed HIP to elevate density of printed IN 718 after normal sintering. The density rose up to 6% after performing HIP at 1210 °C and 206 MPa for three hours which brought about a pore-free density of IN 718.

Another technique to increase density is to infiltrate either a second metal or the same metal. After the debinding step and prior to infiltration, partial sintering of the printed parts is required to avoid distorting the geometry of the parts. The common way of infiltration is to put a printed part in contact with low melting point materials and then heat up the assembly to the temperature between the melting point of the infiltrant and the part. The infiltrant then melts and seeps through the porous structure of the parts due to capillary forces. For instance, inkjet 3D printed stainless steel parts were infiltrated with Cu-based alloys (i.e. typically bronze) to produce fully dense parts with little shrinkage, e.g. ~1%, from the as-built parts [20]. Despite the comparatively low hardness of the infiltrated parts compared with normal tool steels (i.e. 30 HRC versus 40 HRC and above), the infiltrated parts demonstrated excellent wear performance and capability to be welded or electroplated [50]. It is worth noting that in the plastic injection molding industry, such an additive manufactured metal tool provides the possibility of creating conformal cooling channels, thereby reducing cycle times and enhancing the parts quality [50].

In addition to infiltration of the same metal, transient liquid-phase infiltration enables densification of inkjet 3D printed metal skeletons with negligible dimensional change [51]. This method follows the fact that carbon can act as a melting point depressant to provide differences in the melting point of steels. In addition to carbon, silicon or other elements can serve as a melting point depressant in infiltrants. The melting point depressant content in the final parts then reach equilibrium during the homogenization stage in which an additional heat treatment with significant holding time above or below infiltration temperature is used to achieve a microstructurally homogeneous composition throughout the infiltrated skeletons. Fig. 5.5 represents the evolution of Ni–Si materials fabricated by this technique [51]. Fig. 5.5a shows remarkable microscopic variations in Ni-Si composition after holding for one hour at 1180°C. The material became homogeneous with further holding at 1200°C for five hours (see Fig. 5.5b). However, localized pockets of residual porosity corresponding to the original locations of powder particles remained in the microstructures. These porosities sank largely in the sample heat treated at the same temperature for 12 hours as can be seen in Fig. 5.5c. Minimizing erosion of the skeletons and maximizing homogeneity in the microstructure of the infiltrated skeletons are of prime importance in this technique. The fabrication of fully dense D2 tool steels is a good example of inkjet 3DP followed by transient liquid-phase infiltration [52]. To attain the target composition of D2 tool steel, 420 stainless steel powder was used to print skeletons. The skeletons were then sintered at 1300 °C to obtain the relative density of 56%. After adjusting the chemical composition of infiltrant to obtain the target composition, infiltrations were carried out at temperatures between 1270°C and 1300°C. Finally, interconnected carbide films surrounding isolated matrix material grains of the infiltrated sample were partially dissolved as a result of austenitizing heat treatment. Relative density of the fabricated D2 steel was measured to be 99.1% to 99.6%. The average values of hardness for the infiltrated samples were 61.7 HRC and 58.8 HRC in as-quenched conditions and quenched and tempered conditions, respectively. Higher temperature of homogenization and repeated applications of austenitizing heat treatment followed by quenching allowed more carbides to be dissolved and brought about grain refinement. According to Kernan et al. [53] the transient liquid-phase infiltration method is capable of being used as a post-processing approach to densify a broad variety of steel skeletons fabricated by either rapid prototyping or powder metallurgy techniques.

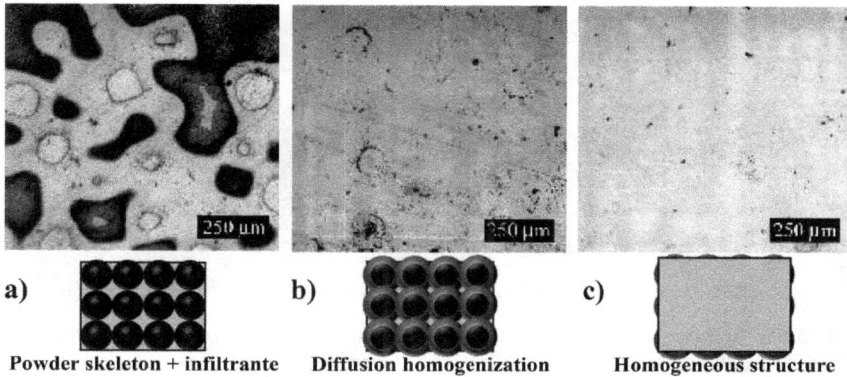

a) Powder skeleton + infiltrante   b) Diffusion homogenization   c) Homogeneous structure

Figure 5.5 Homogenization of Ni–4 wt% Si material fabricated by transient liquid-phase infiltration after: a) heat treatment for 1 hours at 1180 °C, b) 5 hours at 1200 °C, and c) 12 hours at 1200 °C (adapted from [51]).

5.2.2 Shrinkage

The majority of dimensional changes in printed parts arise because of the sintering step. The capillary force, which usually accounts for the main driving force of particles' rearrangement in sintering process, leads to isotropic shrinkage behavior [54]. In contrast, gravitational force, which also acts on particles during sintering, causes non-uniform shrinkage in x, y and z-directions [22, 55, 56]. Preferred pore orientation is another origin of shrinkage anisotropy which exists and acts even on loose powder samples [46].

In essence, the percentage of shrinkages in z-direction is higher compared to the other directions [1, 8, 22, 38]. When the binder materials are burnt during the debinding step, metal particles are pulled down by gravitational force to fill out the void volumes. Furthermore, gravitational force acts on particles to push them down in the course of particle placement during the sintering process. Increasing sintering temperature and time favor the effect of gravitational force, leading to more compaction along the z-direction. Due to the force of gravity, an inkjet 3D printed part may distort or collapse after binder removal and before reaching sintering temperature since the binder free part may not withstand its own weight. These occurrences are shown in Fig. 5.6. One way to mitigate such an issue is to use settering. As mentioned in section 4.4.3, settering makes it possible to prevent gravitational slumping and other forms of parts' distortion. However, settering is not always advantageous since the printed parts tend to contract in the course of sintering, whereas the settering materials have a tendency to withstand the shrinkage.

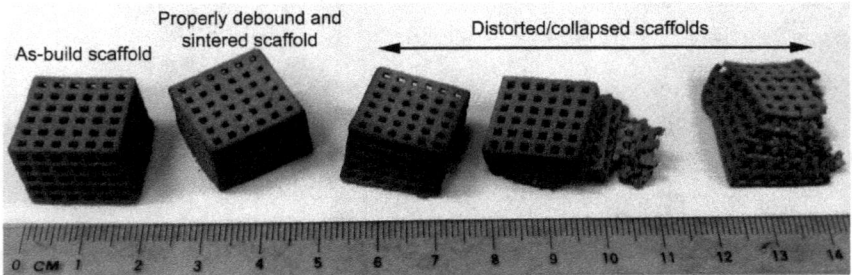

Figure 5.6 Distortion or collapse of Ti scaffolds printed by the SJ on dry mixed feedstock method after debinding and sintering due to the impact of gravitational force.

Values of shrinkage for 3D printed parts, broadly speaking, vary from 3% to 80% [1, 4, 5]. The latter value is for printed porous structures by design. Powder specifications, printing and post-printing related parameters, and sintering profiles are the parameters affecting the shrinkage level of inkjet 3D printed parts. Fig. 5.7 shows how binder concentration can affect the percentage of shrinkage and dimensional accuracy of inkjet 3D printed parts after sintering. As can be seen, for the feedstock containing 5 wt% binder, shrinkage was quite uniform, and distortion was very low. With increasing the percentage of binder materials, shrinkage increased and distortion became very high.

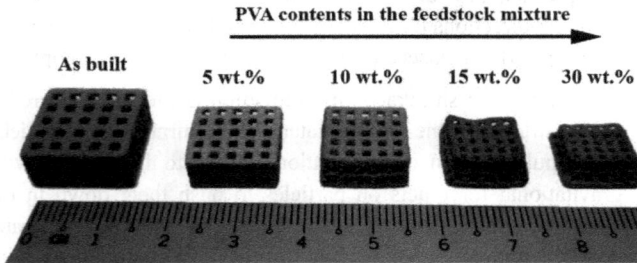

Figure 5.7 The effect of binder percentages on shrinkage and dimensional accuracy of titanium parts printed by the SJ on dry mixed feedstock method after sintering at 1300°C.

Dilatometry results for fine and coarse powder demonstrated that shrinkage values during sintering process are highly dependent on sintering temperature [31]. With rising sintering temperature, shrinkage value increases for both fine and coarse powder. Some researchers have subscribed to the view that smaller particles show larger percentages of shrinkage [57, 58], however, it is not possible to make a general comment on the relation between shrinkage and particle size due to variation in the rate of shrinkage with temperature, according to the observation of the others [31]. Fig. 5.8 displays shrinkage variations of three different particle sizes as a function of temperature for inkjet 3D printed Inconel 718 parts via the liquid BJ on the powder bed method whereby samples sintered at higher temperatures demonstrated higher percentages of shrinkage, irrespective of particle size [16]. As can be seen in the figure, the finest particles size showed the greatest shrinkage value for all studied temperatures, whereas the shrinkage plots for 21 µm and 70 µm particle sizes intersect at 1250°C, indicating different shrinkage behavior as function of particle size and temperature.

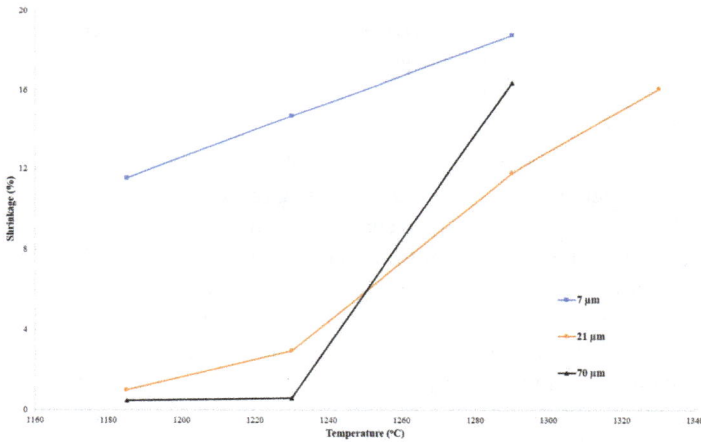

Figure 5.8 Variation in shrinkage of inkjet 3D printed Inconel 718 samples with three different particle sizes as a function of temperature (adapted from [16]).

Controlling the precise dimensions of the final parts and maintaining isotropic levels of shrinkage in all directions are difficult tasks to achieve. Isotropic shrinkage can be resolved by augmenting shrinkage compensation percentages to the constructs with simple geometry at the CAD stage of a model. However, anisotropic shrinkage and creep deformation under structures' self-weight reduce dimensional accuracy, particularly for higher sintering temperatures [59]. One approach to alleviate dimensional changes is to disperse nanoparticles into the printed parts. The dispersion can be conducted either by direct inkjet printing [30] or by infiltration of particles slurry after printing [59]. It is well

109

known that nano-sized particles sinter at much lower temperature compared to micron-sized particles. When these nanoparticles dispersed among micron-sized powder sinter, the neck sizes between particles enlarge and stress in the necks zone reduces. Meanwhile, the bulk of parts made by micron-sized particles would not sinter or sinter to lesser extent during such a heat treatment. Making use of nanoparticles, the interparticle necks gain strength with the lesser center-to-center motion of micron-sized particles which form the bulk of part. Therefore, the overall shrinkage reduces. It should be noted that nanoparticles can be selected to be compatible with the parts' functionality. Hence the final properties are not degraded. In iron-based parts, either iron oxide or iron nanoparticles are ideal candidates as adding several percentages more iron has almost no effect on the final parts' composition. For instance, adding iron nanoparticles into metal skeletons by infiltrating the slurry of particles declines sintering shrinkage up to 50% and creep deflections up to 95% [59]. Moreover, higher final density, lesser distortion, better dimensional stability at sharp corners, greater ductility, and higher tensile strength are the other advantages associated with the presence of nanoparticles together with micron-sized particles [30].

The other approaches of alleviating dimensional changes are to take advantage of metal salt as a binder material (i.e. the salt solutions jetting on powder bed method) [60, 61] and to use binder materials containing a melting point depressant element [62]. If binder materials consist of the components that sinter below the sintering temperature of the printed parts, these components will form interparticle necks. Silver nitrate ($AgNO_3$), for example, was deposited as an aqueous solution into pure Ti [39]. After sintering at 450 °C, which is far below the typical sintering temperature of pure Ti (higher than 1100 °C), silver reduced from the nitrate, forming a well-defined neck between the Ti particles and thus providing dimensional stability. Furthermore, the printed Ti-5Ag alloys demonstrated much higher hardness value as well as better corrosion resistance in comparison with pure titanium [63]. Using this concept, it is not only possible to significantly decrease the shrinkage values, it is also likely to eliminate the standard sintering step [60].

5.2.3 Resolution and Surface Roughness

Resolution is the smallest possible feature size or the finest printable size. In essence, the attainable resolution by inkjet 3DP is governed by the interaction between powder feedstock and ink droplets. The smaller the size of the projected ink drop and powder feedstock, the finer the resolution becomes. Best achievable resolution is twice the size of feedstock particle. From practical standpoint, all printing parameters can impact on resolution to some extent since they are directly or indirectly involved in determining the powder bed feature. Drop placement accuracy is another determining factor. The precision of projecting the drop into the powder bed is determined by the accuracy of movement of the print heads, by the inherent reliability of ink ejections by each nozzle of the print heads, and by the behavior of the droplets in flight which may be dependent upon the presence of other adjacent drops due to electrostatic and aerodynamic impacts [64]. Nonetheless, these operational related parameters are considered as a black box from material developments point of view. Considering the imposed limitations by these parameters, the practical resolution better than tens of micrometres is hard to achieve in inkjet 3D printed parts, which is poorer than some other AM techniques such as SLM or SLA [65, 66].

Similar to the resolution, surface roughness of inkjet 3D printed parts is dominated by the interaction between powder feedstock and ink drops. Essentially, lateral and vertical spreading of ink dictates attainable surface roughness. As mentioned in chapter 4, the spreading behavior is directly influenced by the ink saturation level, layer thickness, and drying situations. Additionally, other printing parameters such as roller spread speed and powder-to-feed ratio have indirect effects on surface finish as a result of their influence on the quality of the powder bed. In this regard, considering the effect of four printing parameters into accounts, Taguchi design of experiment investigations into the end product properties of liquid BJ on powder bed printed 420 stainless steel indicated that the lowest surface roughness (i.e. 12.8 µm) is a result of an optimal combination of the process parameters [67]. Furthermore, among the investigated parameters, namely, layer thickness, drying time, heater power ratio, and ink saturation level, within their studied ranges for each parameter, layer thickness was the most influential one. One plausible reason for the importance of layer thickness on surface roughness is associated with the effect of layer thickness on the dimensional resolution as illustrated in Fig. 4.13. This effect, also known as the stair-stepping phenomenon [68], indicates that the smaller layer thickness causes the part to be produced with finer series of slices and, in turn, the finer surface roughness can be obtained, especially for parts with greater geometrical complexity (see Fig. 4.13). Another way to alter the spreading of ink and improve surface roughness is to modify the characteristics of powder feedstock. It demonstrated that using bimodal powder distribution improved surface roughness which stemmed from the preferential rearrangement of the powder bed while interacting with the drops [69]. Moreover, fine-tuning the relative concentration of fine and large particles dramatically enhanced the surface roughness of inkjet 3D printed parts.

5.2.4 Porosity Related Properties

Elastic moduli of metallic orthopedic implants are approximately 6–12 times greater than that of cortical bone (i.e.7–30 GPa) [70]. This mismatch may cause the stress shielding effect to occur. To alleviate this effect, researchers have introduced pores into metallic structures that significantly lower the modulus of the implants [71, 72]. Moreover, porous structures not only stimulate bone ingrowth, but also allow cell proliferation and differentiation throughout the bulk of the implant materials, thereby enabling enhanced bonding between implants and bone [73, 74]. The SJ on dry mixed feedstock method has been used extensively to fabricate metallic scaffolds with dual porosity features, namely:

i.   macropores or pores by design and

ii. micropores or pores by process [1-8].

The overall porosity percentage of structures is a contribution of these micropores and macropores so that the porosity level higher than 80% is readily achievable. Fig. 5.9 shows a bimodal pore size of Ti scaffold with porosity percentage above 80 %.

Figure 5.9 Inkjet 3D printed and sintered Ti scaffolds containing two types of pores.

Similar to the SJ on dry mixed feedstock method, the idea of space filler can be employed using the liquid BJ on powder bed method to obtain porous structures. It is possible to mix fugitive materials with metal powder in the powder feedstock preparation stage. The fugitive materials act as space holder during the build cycle and then they evaporate during sintering to leave void spaces behind. For instance, PA 2202 powder (heat stabilized, pigment-filled polyamide 12 grade) with an average particle size of 55 μm was mixed with 316L powder having $D_{90} < 22$ μm with the ratio of 1:3 or 1:4 [22]. The liquid BJ on powder bed method was used for inkjet 3DP of the prepared powder feedstock. After debinding and space holder burnout at 800 °C, the samples were sintered at 1360 °C for one hour. Fig. 5.10 shows cross-sectional micrographs of sintered samples with zero, 25 %, and 33% of the space filler mixtures, providing 7 to 36 % porosity. Such a porous structure can be used to control heat transfer. Furthermore, other materials can be infiltrated into the resulting porous structures to introduce new functionalities to the structure.

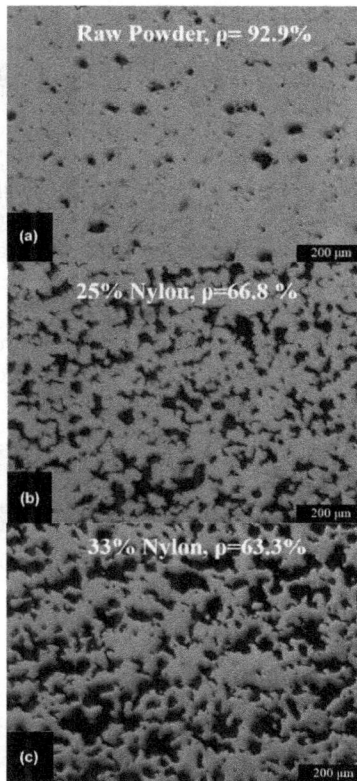

Figure 5.10   Cross-sectional optical micrographs of the samples made of different feedstock materials: a) 0 % space filler with 100% 316L powder, b) 25% space filler with 75% 316L powder, and c) 33% space filler with 67% 316L powder.

113

Given the advantages of inkjet 3DP, either rectangular [4] or cylindrical [8] macropores designed via CAD models are printable. The minimum printable size of pores by design is five times of feedstock's particle size [75]. However, removing unbound powder from deep inside structures without damaging as-built parts is a difficult task [24]. These designed pores may be as small as 500 μm after sintering, demonstrating the potential of inkjet 3DP to produce a user-controlled porous structure [3, 76]. The value of porosity by design can be correlated to characteristics of rectangular pores by the following equation:

$$\text{Porosity percentage} = \frac{P^2(P+3W)}{(P+W)^3} \tag{5.1}$$

Where, P is pore width and W is the strut width between each pore. Pores by process are classified according to their sizes, namely the larger micropores and the smaller micropores. The former one is generated by the burning of the binders in the debinding step in which the burnt binders leave a void volume behind, especially for constructs printed by the SJ on dry mixed feedstock method. The smaller micropores arise from remnants of packing gaps between metal particles after the sintering process. The size of micropores is dictated by post-printing conditions (i.e. sintering profile), size and content of the binder materials as well as mean diameter of the metal particles. The finer the particle size, the smaller the pores between particles. The sintering temperature does not affect the percentage of pores by design, whereas sintering at higher temperature reduces the number of open pores as well as the degree of pores' interconnectivity. For instance, when the sintering temperature is increased to obtain higher relative density, e.g. 85–90%, the number of open pores is almost reduces to zero [12]. The typical sizes of pores by this process are between 20 μm and 70 μm, more than 90% of which could be interconnected [3]. In addition, the size of pores by this process can be tailored in accordance with the applications. For example, it is advisable to create pores in the 100 to 300 μm size range with this process [21] which is a desirable range for tissue engineering [77].

As a consequence of the rough surface finish, there is no need for further surface modification treatments to enhance the material-cell interactions for inkjet 3D printed scaffolds [21]. Post-printing processes, binders content, and shapes of metal particles determine the total surface roughness. A higher percentage of binder materials makes the surface finish rougher due to the lower packing density and the poorer stacking of the powder. It can be seen in Fig. 5.11 that particle shape has a considerable impact on the resultant pores' roughness: the pathways between spherical particles are smoother than those between non-spherical particles.

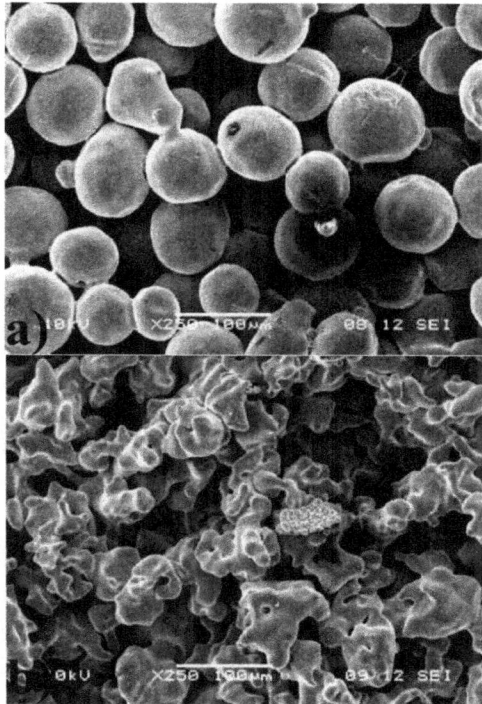

Figure 5.11   Scanning electron micrographs of pores roughness of printed 316 L stainless steel with: a) spherical 45–90 µm powder and b) non-spherical 45–90 µm powder [12].

As stated in the first chapter, the main drawback of inkjet 3DP technology is associated with low density and porous structure of printed parts which deteriorate mechanical properties. This is because pores not only increase local stress due to the stress concentration effect but also act as preferential locations for crack initiation and propagation. Pores characteristics in terms of their size, shape and distribution are crucial factors in evaluating mechanical properties of inkjet 3D printed parts. Optimizing sintering profile remarkably improves the pores' characteristics. Fig. 5.12 shows optical micrographs of inkjet 3D printed IN 625 samples sintered at different temperatures ranging from 1200 to 1290 °C [17]. For the samples sintered at 1200 and 1220 °C, highly interconnected porous structures are discernible. With increasing sintering temperature, interconnected pores with irregular shape progressively alter to separated ones with spherical shape as well as decrease in porosity percentage of the samples. Table 5.2 represents variation of pore size with sintering temperature for inkjet 3D printed IN 625 samples and Fig. 5.13 displays selected SEM images of these samples [18]. As can be seen, pores sizes reduced from 41 to 2.6 μm as sintering temperature rose from 1225 to 1300 °C and accordingly density improved and reached to 99.2%, up from 85%. Thus, the microhardness value as representative of mechanical properties gradually enhanced from 160 HV to more than 200 HV.

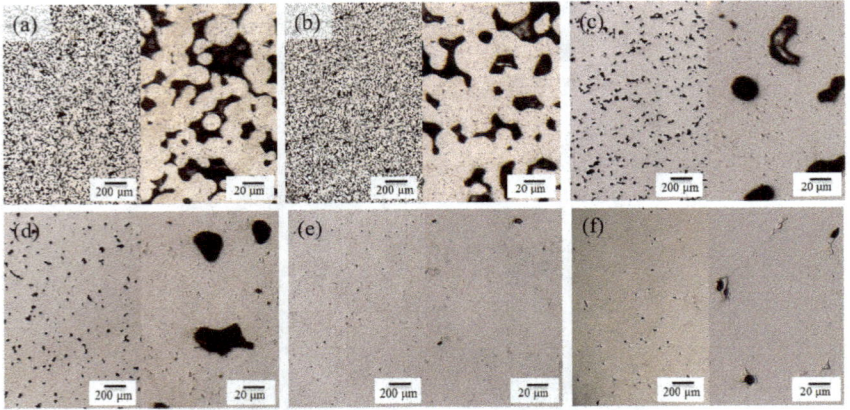

Figure 5.12   Optical micrographs of inkjet 3D printed IN 625 alloys samples after 4 hours sintering at different temperatures: (a) 1200 °C, (b) 1220 °C, (c) 1240 °C, (d) 1260 °C, (e) 1280 °C, and (f) 1290 °C (courtesy – A. Mostafaei [17]).

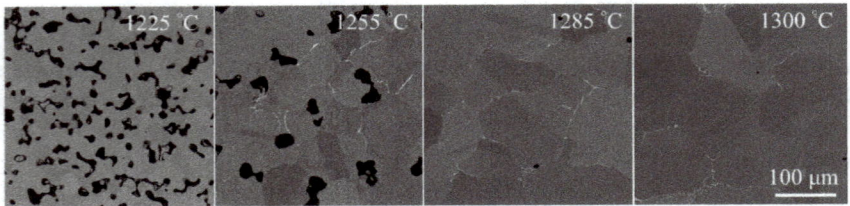

Figure 5.13   SEM micrographs of samples after 4 h sintering at different temperatures (courtesy – A. Mostafaei [18]).

Table 5.2 Variation of pore size for inkjet 3D printed IN 625 samples sintered with different temperature [18].

Pore size (μm)	Sintering temperature (°C)
41 ± 18	1225
36 ± 10	1240
27 ± 13	1255
20 ±4	1270
4.5 ± 1.2	1285
2.6 ± 0.4	1300

5.2.5 Biological Properties

In vitro studies into culturing of adipose tissue-derived mesenchymal stem cells (ASCs) on the 2D dense and inkjet 3D printed Ti are shown in Fig. 5.14. It can be seen that the higher number of cells attached and later proliferated on the inkjet 3D printed samples, as against 2D dense samples. This is attributed to the higher surface roughness and porosity values of the inkjet 3D printed samples [9]. Furthermore, SEM analysis of cell-seeded samples revealed a fibroblast-like morphology of the cells on the inkjet 3D printed samples, while the cells on the 2D Ti exhibited elongated and stressed morphologies. This may be due to the fact that the rough inkjet 3D printed surfaces provide more cell attachment points as compared to the relatively smooth surfaces of 2D dense Ti. In addition, the 3D interconnected pore architecture allows the cells to penetrate into structures, hence further improves the cells viability and proliferation. Fig. 5.15 shows cell culture studies on porous Ti cages which demonstrated that the surfaces of printed implants were fully covered with living cultured cells, and the cells could attach and proliferate well with a minority amount of dead cells. Likewise, healthy elongated and spread morphologies of adhered cells were observed on inkjet 3D printed Fe–30Mn and Fe-30Mn-1Ca scaffolds [24, 25]. All these results strongly suggest the promise of inkjet 3D printed structures as suitable substrates to support cell adherence, proliferation, and osteogenic differentiation.

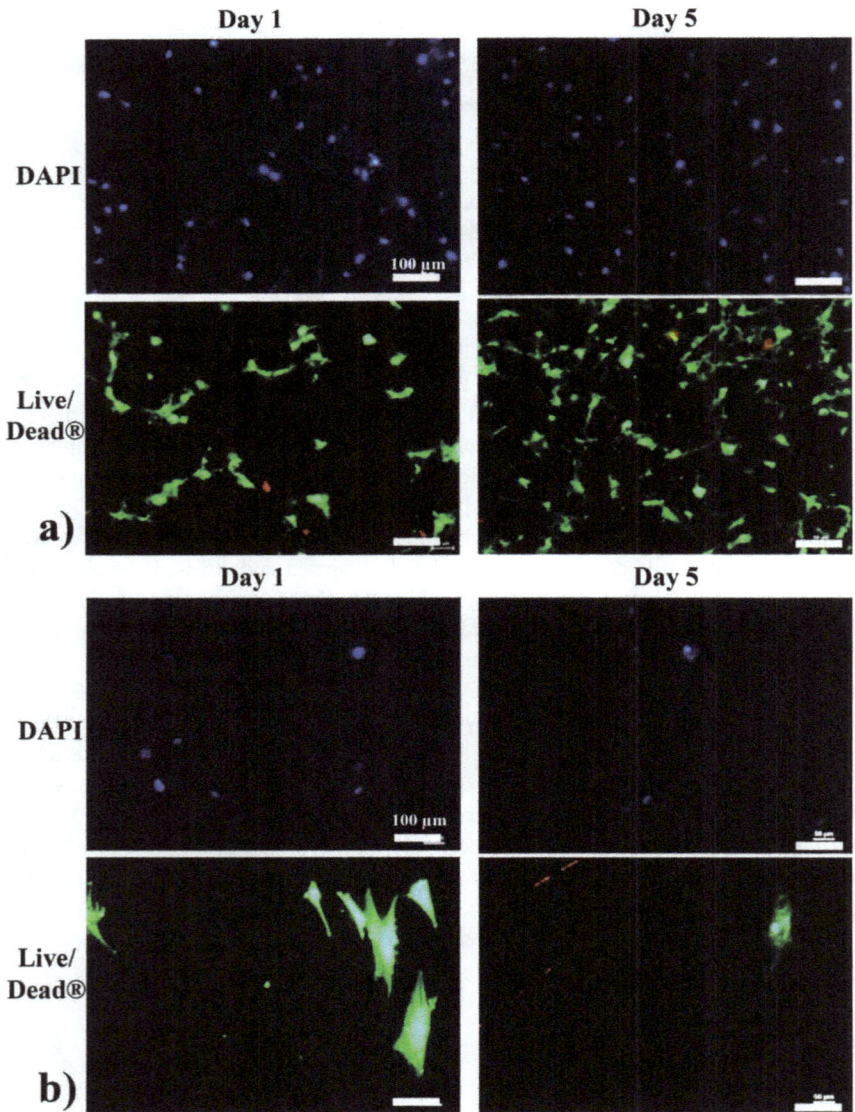

Figure 5.14 DAPI and LIVE/DEAD® staining of ASCs cultured on: a) inkjet 3D printed titanium and b) 2D dense titanium samples at various time points (adapted from [9]).

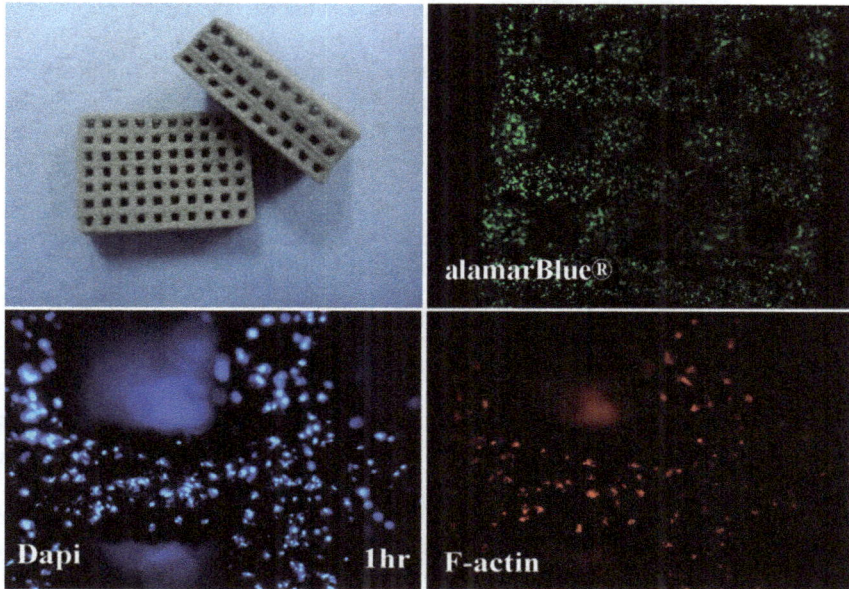

Figure 5.15 Cell proliferation on 3D printed titanium cages.

5.2.6 Mechanical Properties

It would be rational to assume the inkjet 3D printed parts as a homogenous porous structure when it comes to mechanical property evaluation. To predict compressive strength of porous metals with homogenous pores distribution, the model of Gibson and Ashby can be applied as following [1, 78]:

$$\frac{\sigma}{\sigma_s} = C \left(\frac{\rho}{\rho_s} \right)^{3/2} \tag{5.2}$$

Where, σ is the strength of the bulk metal, σ_s is the strength of a porous metallic structure, C is a proportionality constant, and ρ and ρ_s are the density of the porous and bulk metal, respectively. In this model, compressive strength only depends upon the relative density. Nonetheless, the difference between experimental results and predicted values was large when this model was used for Ti porous structure [79]. Consider the case of two inkjet

3D printed Ti samples which have the same porosity level of 36% but produced differently in terms of powder size and sintering conditions. The model predicted the same strength although the experimental compressive strength's values were 144 MPa and 408 MPa [79]. Given that higher sintering temperature, longer sintering time or finer powders size results in greater compressive strength, Basalah et al. [1] found that the ratio of interparticle neck size to particle size have a major role in the mechanical strength of the printed porous structures. Thus, both the sinter neck size and the relative density must be taken into account.

The elastic modulus of a porous structure is correlated to the bulk modulus, wall thickness and pore size by the following equation [3].

$$E_p = E_b \times \left(\frac{W}{W+P} \right)^2 \tag{5.3}$$

Where, E_p and E_b are the modulus of porous structure and bulk material respectively, W is the strut width between each pore and P is the pore width. Using equations 5.1 and 5.3, Fig. 5.16 shows the variation of the relative modulus (E scaffold / E bulk) of the inkjet 3D printed porous structures with randomly selected pore size and wall thickness values as a function of corresponding porosity percentages. As can be seen from Fig. 5.16, there is a strong correlation between the elastic modulus of a printed porous structure and its level of porosity indicating that the stress shielding effect may be alleviated remarkably using the inkjet 3D printed scaffolds. Fig. 5.17a displays a material-property chart for natural bone and several natural/synthetic materials that can be utilized as bone and tooth replacements. As can be seen from the figure, metallic materials and bone are apart from each other. However, taking advantage of the pores by process makes it possible to adjust the properties of Ti alloys in a manner that the properties of inkjet 3D printed Ti match to that of the cortical bone as illustrated in Fig. 5.17 b. Similarly, it is plausible to attain the properties of cancellous bone by introducing the pores by design in inkjet 3D printed Ti scaffolds. Inclusive comparisons between mechanical properties of dense Ti, inkjet 3D printed Ti, cortical bone, and cancellous bone can be made using provided data in Table 5.3. As can be seen from the table, mechanical properties of inkjet 3D printed Ti parts can be altered by fine-tuning porosity levels and sintering profiles.

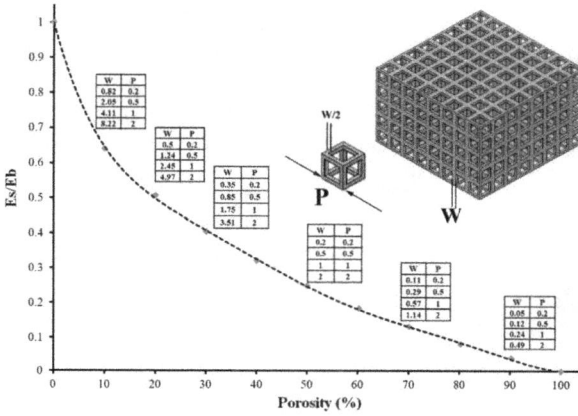

Figure 5.16 Variation of the relative modulus (E scaffold / E bulk) of inkjet 3D printed porous structures as a function of the corresponding porosity percentages [3].

Given the importance of materials behavior under tension, tensile properties of an inkjet 3D printed metal in comparison with those of conventional manufacturing processes can be taken into account as a reliable performance indicator of the inkjet 3DP technology for metals. Comparisons between tensile behavior of inkjet 3D printed Inconel 625 with conventionally processed alloys via either powder metallurgy or casting techniques demonstrated similar mechanical properties in terms of yield and ultimate tensile strength as well as ductility [17-19]. This implies that the inkjet 3DP technology can be considered as a reliable AM technique for metals processing if the as-built parts are optimally sintered.

Mechanical properties of inkjet 3D printed metallic parts with a wide variety of porosity levels and sintering profiles fabricated with different printing methods are summarized in Table 5.3. As can be seen from the table, mechanical properties of the same part can be altered widely to obtain intended properties. Variations in the values of mechanical properties largely stem from different porosity percentages and sintering profiles. In general, it can be concluded from Table 5.3 that the higher sintering temperature and the lower porosity level result in higher mechanical properties. In essence, sintering profile and porosity level are the two dominating factors determining mechanical properties. It should be noted that these two factors are not independent parameters; they may depend on each other as well as the other parameters. Consider the aforementioned case of correlation between sintering profile and the ratio of interparticle neck size to particle size. As another example, the level of porosity in inkjet 3D printed parts is greatly influenced by characteristics of feedstock materials and printing parameters as discussed in previous chapters. A case in point is the effect of layer thickness. A prevalent reason for the failure of inkjet 3D printed parts is the sliding between layers caused by shearing in the core layers and sliding at the interfaces between the layers [7]. Thus, an effective measure to minimize risks of the sliding is to reduce the layer thickness.

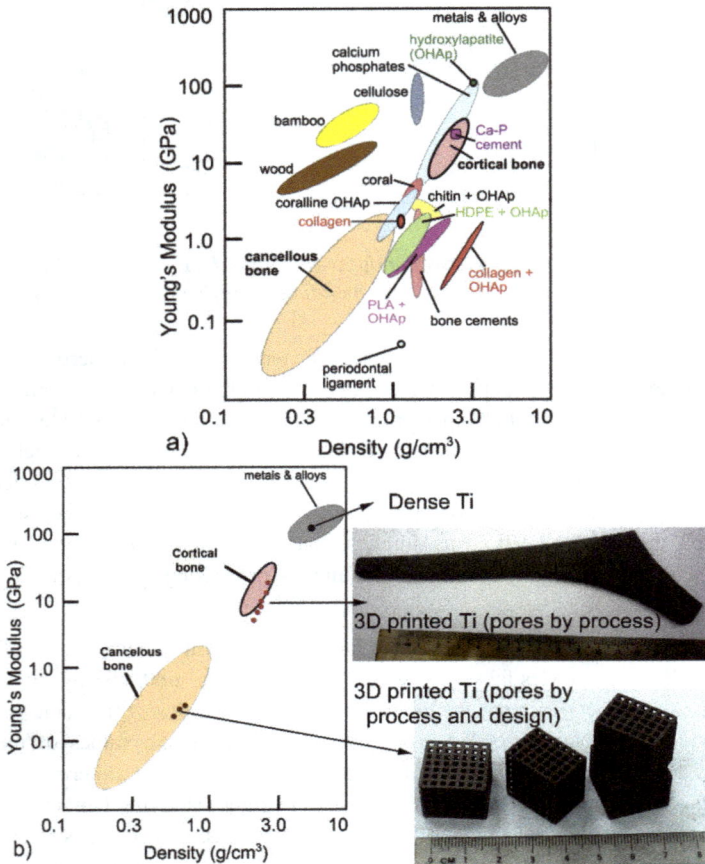

Figure 5.17 a) The plot of Young's modulus versus density (materials property chart): a) for natural bone and several other synthetic/natural materials that can be utilized as bone or tooth replacements [80] and b) for cancellous bone, cortical bone, and inkjet 3D printed Ti scaffolds.

Very few studies have been carried out to explore the impacts of inkjet 3DP parameters on mechanical properties of metallic parts either in as-built or sintered conditions. In this regard, the investigation into the effect of inkjet 3DP parameters using a design of experiments approach on flexural strength of 316L stainless steel samples provides fresh insights [81]. In this study, the liquid BJ on powder bed method was used to fabricate samples with 27 different combinations of printing parameters. These combinations were created from the variations in four parameters, namely, ink saturation level, roller spread speed, feed-to-powder ratio, and layer thickness, while each parameter varied in three levels. All samples underwent similar post-printing steps in terms of the debinding and sintering profile. The results indicated that printing parameters have a remarkable influence on the mechanical properties whereby flexural strengths fluctuated between 38.5 and 90 MPa (more than 130 %), implying the prime importance of judicious selection of printing parameters to achieve desired properties in the end products. Furthermore, in order to identify the individual effect of each parameter on flexural strength, a statistical analysis was carried out which revealed that all parameters can alter the flexural strength to some extent. However, among the parameters and within the ranges studied in this research, ink saturation level and feed-to-powder ratio were the most influential parameters on the flexural strength of the samples. Far more investigations are required to systematically determine the influence of each individual parameters as well as the possible interactions between them. Design of experiments approach such as Taguchi, response surface methodology, and neural network methods could be used for such exploration to minimize the number of experiments required.

Table 5.3 Mechanical properties of metallic parts fabricated by inkjet 3DP.

Material systems	Porosity percentage (%), Porosity by design*	Mechanical properties								Sintering profile	Ref
		Modulus of elasticity (GPa)	Yield strength (MPa)	Ultimate tensile strength (MPa)	Compressive strength (MPa)	Fracture strength (MPa)	Elongation (%)	Fracture toughness ($MPa.m^{1/2}$)	Hardness		
Pure Ti	31-43, None	-	-	-	56-509	-	-	-	-	1100 °C and 1400 °C for 1-3 hours	[1]
Pure Ti	19.2-64.3, None	0.76-1.72	38.3-107.2	-	-	-	-	-	161-286 (Hv)	900-1350 °C	[2]
Pure Ti	32.2-52.7, None	0.9-8.6	-	-	-	69.1-245.7	-	2.4-16.5	18.1-33.5 (HRb)	1250-1370 °C for 2 and 5 hours	[3]
Pure Ti	83-85.9, Y	0.31-0.33	-	-	-	-	-	-	-	1000-1350 °C for 2 hours	[4]
Pure Ti	-, Y	4.8-13.2	-	-	166.9-455.1	-	-	-	-	1250-1350 °C for 2 hours	[5]
Pure Ti	36-40, None	1.27-2.12	68.2-80.2	-	-	-	-	-	-	900-1300 °C	[6]
Pure Ti	41.1-65, None	5.9-34.8	-	-	81.3-218.6	-	-	-	115.2-182.8 (Hv)	1200-1400 °C	[7]
Composite of 80%Ti-20%HA	41.1-65, None	-	-	-	184.3	-	-	-	-	1200 °C	[34]
Pure Ti	33.6-35.5, Y	3-3.4	148.8-156.1	306.6-414.5	-	-	4.9-5.8	-	-	1400 °C	[8]
Cortical bone	-	7-30	-	50-150	130-180	-	0.72-1.45	2-12	32 (HRb)	-	[70, 82]
Cancellous bone	-	0.1-4.5	-	2-20	2-12	-	-	0.1-0.8	-	-	[9, 82]
Dense pure Ti	0%, None	105	-	79-345	-	-	-	50	80 (HRb)	-	[9]
Ti-5Ag	1.5-6.3%, None	-	-	522-612	-	-	-	-	429-700 (Hv)	1150 and 1300 °C for 1 hour	[39]
Inconel 625	21-0.4 %, None	-	327	-	-	-	41	-	107-237(Hv)	1200-1300 °C for 4 hours	[17]
Inconel 625	< 1%, None	-	287 and 376	386 and 644	-	-	11.6 and 47	-	191and 213 (Hv)	1270 and 1285 °C for 4 hours	[18]

Material systems	Porosity percentage (%), Porosity by design*	Mechanical properties								Sintering profile	Ref
		Modulus of elasticity (GPa)	Yield strength (MPa)	Ultimate tensile strength (MPa)	Compressive strength (MPa)	Fracture strength (MPa)	Elongation (%)	Fracture toughness (MPa.m$^{1/2}$)	Hardness		
Inconel 625	0.4 %, None	-	327, 311[1], 373[2] and 392[3]	612, 587[1], 647[2], and 697[3]	-	-	41, 45[1], 36[2], and 30[3]	-	237,284[1], 320[2], and 333[3] (Hv)	1280 °C for 4 hours	[19]
Inconel 625 after aging treatment	<1%, None	-	298 and 394[4]	405 and 718[4]	-	-	7.5 and 29[4]	-	225 and 327[4] (Hv)	1270 and 1285 °C for 4 hours	[18]
420 stainless steel	-, None	-	455	680	-	-	-	-	26-30 (HRc)	1120 °C	[11]
17-4PH stainless steel	5%, None	-	696 and 1006[5]	765 and 1110[5]	-	-	-	-	205 and 345[5] (Hv)	1330 °C for 3 hours	[11]
420 stainless steel	0%, None	-	-	-	-	-	-	-	35 and 61.7 (HRc)	Printing skeleton+ infiltration	[52]
420 stainless steel	29.3-47.7, None	-	-	131-215	-	-	-	-	-	1100-1300 °C for 1.5-20 hours	[13]
316L stainless steel	1.3-38.3, None	-	-	309-518	-	-	21.3-61.9	-	-	1200-1405 °C for 1.5 hours	[12]
Fe-30 Mn	-, None	32.5	106	115.5	-	-	0.73	-	-	1200 °C for 3 hours	[24]
Cu	14.5-27.8, None	-	-	116.7	-	-	-	-	-	1080-1090 °C for 2-4 hours	[26]
Ag	12.9-11.1, None	-	-	46-55	-	-	-	-	-	850 °C for 20 min	[30]

*Y means there were pores by design in parts, while None means that there was no pore by design in parts

1, 2 and 3: sintered and[1] solutioned (1150 °C, 2 h),[2] Aged (745 °C, 20 h), and[3] Aged (745 °C, 60 h)

4: sintered followed by age treatment at 745 °C for 20 h

5: sintering followed by heat treatment

Summary

Various inkjet 3DP methods have been utilized to fabricate parts made of monolithic alloys, especially Ti, stainless steel, and nickel based superalloys, as well as composites and functionally graded materials. Printing methods, powder feedstock specifications, and sintering profile have significant impact on the final density of the inkjet 3D printed metallic parts.

In addition to modification of the powder feedstock by adding sintering aid elements, various post-printing processes such as performing HIP and infiltrating molten metals are applicable to achieve near full dense parts.

Among several parameters influencing resolution of inkjet 3D printed parts, ink droplet size and powder particle size are the most relatively straightforward parameters to change. Surface roughness of inkjet 3D printed parts is governed by lateral and vertical spreading of ink, printing parameters and powder feedstock features. The percentage of shrinkage associated with the post-printing process of inkjet 3D printed parts is dependent on many factors ranging from the features of powder feedstock to the implementation of the post-printing process. It is plausible to mitigate shrinkage by making use of techniques such as infiltration, dispersion of nanoparticles, metal salt binders, and a melting point depressant element.

With the ability to control the size and morphology, highly porous metallic scaffolds with interconnected porous structure fabricated by inkjet 3DP allow cell proliferation and differentiation throughout the bulk of scaffolds.

Mechanical properties of inkjet 3D printed metals can be tailored to suit specific requirements. The properties are principally governed by porosity level and sintering profile. However, these factors also depend on other parameters such as printing parameters.

References

1. Basalah, A., et al., Characterizations of additive manufactured porous titanium implants. J Biomed Mater Res Part B, 2012. 100B: p. 1970-1979. https://doi.org/10.1002/jbm.b.32764

2. Gagg, G., E. Ghassemieh, and F.E. Wiria, Effects of sintering temperature on morphology and mechanical characteristics of 3D printed porous titanium used as dental implant. Materials Science and Engineering: C, 2013. 33(7): p. 3858-3864. https://doi.org/10.1016/j.msec.2013.05.021

3. El-Hajje, A., et al., Physical and mechanical characterisation of 3D-printed porous titanium for biomedical applications. Journal of Materials Science: Materials in Medicine, 2013. 25(11): p. 2471-2480. https://doi.org/10.1007/s10856-014-5277-2

4. Maleksaeedi, S., et al., Toward 3D Printed Bioactive Titanium Scaffolds with Bimodal Pore Size Distribution for Bone Ingrowth. Procedia CIRP, 2013. 5: p. 158-163. https://doi.org/10.1016/j.procir.2013.01.032

5. Wiria, F.E., et al., Printing of Titanium implant prototype. Materials & Design, 2010. 31, Supplement 1: p. S101-S105. https://doi.org/10.1016/j.matdes.2009.12.050

6. Gagg, G., E. Ghassemieh, and F.E. Wiria, Analysis of the compressive behaviour of the three-dimensional printed porous titanium for dental implants using a modified cellular solid model. Proceedings of the Institution of Mechanical Engineers, Part H: Journal of Engineering in Medicine, 2013. 227(9): p. 1020-1026. https://doi.org/10.1177/0954411913489802

7. Xiong, Y., C. Qian, and J. Sun, Fabrication of porous titanium implants by three-dimensional printing and sintering at different temperatures. Dental Materials Journal, 2012. 31(5): p. 815-820. https://doi.org/10.4012/dmj.2012-065

8. Basalah, A., S. Esmaeili, and E. Toyserkani, Mechanical properties of additive-manufactured porous titanium bio-structures with oriented macro-scale channels. The International Journal of Advanced Manufacturing Technology, 2015: p. 1-8.

9. Guneta, V., et al., Three dimensional printing of titanium for bone tissue engineering applications: A preliminary study. Journal of Biomimetics, Biomaterials & Tissue Engineering, 2014. 21: p. 101-115. https://doi.org/10.4028/www.scientific.net/JBBBE.21.101

10. Hong, S.B., et al., A new Ti-5Ag alloy for customized prostheses by three-dimensional printing (3DP). J Dent Res, 2001. 80(3): p. 860-3. https://doi.org/10.1177/00220345010800030301

11. Carreño-Morelli, E., et al., Three-Dimensional Printing of Stainless Steel Parts. Materials Science Forum, 2008 591-593: p. 374-379. https://doi.org/10.4028/www.scientific.net/MSF.591-593.374

12. Verlee, B., T. Dormal, and J. Lecomte-Beckers, Density and porosity control of sintered 316L stainless steel parts produced by additive manufacturing. Powder Metallurgy, 2012. 55(4): p. 260-267. https://doi.org/10.1179/0032589912Z.00000000082

13.   Yu Zhou, S.C. Siw, and C.I. Garcia, Microstructural and Porosity Characterization During the Sintering Process Of 420 Stainless Steel Samples Using Powder Based 3D Printing, in TMS. 2014, TMS. p. 1975 - 1984.

14.   Turker, M., D. Godlinski, and F. Petzoldt, Effect of production parameters on the properties of IN 718 superalloy by three-dimensional printing. Materials Characterization, 2008. 59(12): p. 1728-1735. https://doi.org/10.1016/j.matchar.2008.03.017

15.   Sicre-Artalejo, J., et al., High-density inconel 718: Three-dimensional printing coupled with hot isostatic pressing. The international journal of powder metallurgy, 2008. 44 p. 35-43.

16.   Nandwana, P., et al., Powder bed binder jet 3D printing of Inconel 718: Densification, microstructural evolution and challenges. Current Opinion in Solid State and Materials Science, 2017. https://doi.org/10.1016/j.cossms.2016.12.002

17.   Mostafaei, A., et al., Powder bed binder jet printed alloy 625: Densification, microstructure and mechanical properties. Materials & Design, 2016. 108: p. 126-135. https://doi.org/10.1016/j.matdes.2016.06.067

18.   Mostafaei, A., et al., Microstructural evolution and mechanical properties of differently heat-treated binder jet printed samples from gas- and water-atomized alloy 625 powders. Acta Materialia, 2017. 124: p. 280-289. https://doi.org/10.1016/j.actamat.2016.11.021

19.   Mostafaei, A., et al., Effect of solutionizing and aging on the microstructure and mechanical properties of powder bed binder jet printed nickel-based superalloy 625. Materials & Design, 2016. 111: p. 482-491. https://doi.org/10.1016/j.matdes.2016.08.083

20.   Sachs, E., et al., Production of injection molding tooling with conformal cooling channels using the three dimensional printing process. Polymer Engineering & Science, 2000. 40(5): p. 1232-1247. https://doi.org/10.1002/pen.11251

21.   Dourandish, M., D. Godlinski, and A. Simchi, 3D printing of biocompatible PM-materials, in Materials Science Forum. 2007. p. 453-456.

22.   Ziaee, M., E.M. Tridas, and N.B. Crane, Binder-Jet Printing of Fine Stainless Steel Powder with Varied Final Density. JOM, 2017. 69(3): p. 592-596. https://doi.org/10.1007/s11837-016-2177-6

23. Carreño-Morelli, E., S. Martinerie, and J.E. Bidaux, Three-Dimensional Printing of Shape Memory Alloys. Materials Science Forum, 2007 534-536: p. 477-480. https://doi.org/10.4028/www.scientific.net/MSF.534-536.477

24. Chou, D.-T., et al., Novel processing of iron-manganese alloy-based biomaterials by inkjet 3-D printing. Acta Biomaterialia, 2013. 9(10): p. 8593-8603. https://doi.org/10.1016/j.actbio.2013.04.016

25. Hong, D., et al., Binder-jetting 3D printing and alloy development of new biodegradable Fe-Mn-Ca/Mg alloys. Acta Biomaterialia, 2016. 45: p. 375-386. https://doi.org/10.1016/j.actbio.2016.08.032

26. Bai, Y. and C.B. Williams, An exploration of binder jetting of copper. Rapid Prototyping Journal, 2015. 21(2): p. 177-185. https://doi.org/10.1108/RPJ-12-2014-0180

27. Minagawa, K., et al., Application of 3-Dimensional Powder Laminating Fabrication to Metallic Components. Materials Science Forum, 2007. 539-543: p. 2333-2338. https://doi.org/10.4028/www.scientific.net/MSF.539-543.2333

28. Lu, K. and W.T. Reynolds, 3DP process for fine mesh structure printing. Powder Technology, 2008. 187(1): p. 11-18. https://doi.org/10.1016/j.powtec.2007.12.017

29. Lu, K., M. Hiser, and W. Wu, Effect of particle size on three dimensional printed mesh structures. Powder Technology, 2009. 192(2): p. 178-183. https://doi.org/10.1016/j.powtec.2008.12.011

30. John, G.B., D.C. Kevin, and A.K. Howard, Inkjet printable nanosilver suspensions for enhanced sintering quality in rapid manufacturing. Nanotechnology, 2007. 18(18): p. 185701. https://doi.org/10.1088/0957-4484/18/18/185701

31. Dourandish, M., et al., Sintering of biocompatible P/M Co–Cr–Mo alloy (F-75) for fabrication of porosity-graded composite structures. Materials Science and Engineering: A, 2008. 472(1–2): p. 338-346. https://doi.org/10.1016/j.msea.2007.03.043

32. Li, S., et al. Progress toward a denser metal matrix composite using the three dimensional printing method. in Transactions of the North American Manufacturing Research Institution of SME. 2008.

33. Sun, L., et al. Densification and dimensional accuracy of metal matrix composites fabricated by three-dimensional printing. in Transactions of the North American Manufacturing Research Institution of SME. 2009.

34.   Qian, C., F. Zhang, and J. Sun, Fabrication of Ti/HA composite and functionally graded implant by three-dimensional printing. Biomed Mater Eng, 2015. 25(2): p. 127-36.

35.   Godlinski, D. and S. Morvan, Steel Parts with Tailored Material Gradients by 3D-Printing Using Nano-Particulate Ink. Materials Science Forum, 2005. 492-493: p. 679-684. https://doi.org/10.4028/www.scientific.net/MSF.492-493.679

36.   Vlasea, M., et al., A combined additive manufacturing and micro-syringe deposition technique for realization of bio-ceramic structures with micro-scale channels. The International Journal of Advanced Manufacturing Technology, 2013. 68(9): p. 2261-2269. https://doi.org/10.1007/s00170-013-4839-7

37.   Vlasea, M. and E. Toyserkani, Experimental characterization and numerical modeling of a micro-syringe deposition system for dispensing sacrificial photopolymers on particulate ceramic substrates. Journal of Materials Processing Technology, 2013. 213(11): p. 1970-1977. https://doi.org/10.1016/j.jmatprotec.2013.05.011

38.   Wiria, F.E., S. Maleksaeedi, and Z. He, Manufacturing and characterization of porous titanium components. Progress in Crystal Growth and Characterization of Materials, 2014. 60: p. 94-98. https://doi.org/10.1016/j.pcrysgrow.2014.09.001

39.   Hong, S.B., et al., A new Ti-5Ag alloy for customized prostheses by three-dimensional printing (3DP™). Journal of Dental Research, 2001. 80(3): p. 860-863. https://doi.org/10.1177/00220345010800030301

40.   Kakisawa, H., et al., Dense P/M Component Produced by Solid Freeform Fabrication (SFF). Materials Transactions, JIM, 2005. 46(12): p. 2574-2581. https://doi.org/10.2320/matertrans.46.2574

41.   Coble, R.L., Sintering Crystalline Solids. 1. Intermediate and Final State Diffusion Models. Journal of Applied Physics, 1961. 32: p. 787-792. https://doi.org/10.1063/1.1736107

42.   Mostafaei, A., et al., Data on the densification during sintering of binder jet printed samples made from water- and gas-atomized alloy 625 powders. Data in Brief, 2017. 10: p. 116-121. https://doi.org/10.1016/j.dib.2016.11.078

43.   Reid, C.R., Numerical simulation of free shrinkage using a continuum theory for sintering. Powder Technology, 1994. 81(3): p. 287-291. https://doi.org/10.1016/0032-5910(94)02887-7

44.   Hassold, G.N., I.W. Chen, and D.J. Srolovitz, Computer Simulation of Final-Stage Sintering: I, Model Kinetics, and Microstructure. Journal of the American Ceramic Society, 1990. 73(10): p. 2857-2864. https://doi.org/10.1111/j.1151-2916.1990.tb06686.x

45.   Su, H. and D.L. Johnson, Master Sintering Curve: A Practical Approach to Sintering. Journal of the American Ceramic Society, 1996. 79(12): p. 3211-3217. https://doi.org/10.1111/j.1151-2916.1996.tb08097.x

46.   Olevsky, E.A., Theory of sintering: from discrete to continuum. Materials Science and Engineering: R: Reports, 1998. 23(2): p. 41-100. https://doi.org/10.1016/S0927-796X(98)00009-6

47.   Aguirre, I., et al., Effect of nitrogen on supersolidus sintering of modified M35M high speed steel. Powder Metallurgy, 1999. 42(4): p. 353-357. https://doi.org/10.1179/003258999665701

48.   Zhang, Z. and R. Sandstrm, Fe-Mn-Si master alloy steel by powder metallurgy processing. Journal of Alloys and Compounds, 2004. 363(1â€"2): p. 199-207.

49.   Sun, L., et al., Densification and properties of 420 stainless steel produced by three-dimensional printing with addition of Si3N4 powder. Journal of Manufacturing Science and Engineering, Transactions of the ASME, 2009. 131(6): p. 0610011-0610017. https://doi.org/10.1115/1.4000335

50.   Allen, S.M. and E.M. Sachs, Three-dimensional printing of metal parts for tooling and other applications. Metals and Materials International, 2000. 6(6): p. 589-594. https://doi.org/10.1007/BF03028104

51.   Lorenz, A., et al., Densification of a powder-metal skeleton by transient liquid-phase infiltration. Metallurgical and Materials Transactions A: Physical Metallurgy and Materials Science, 2004. 35 A(2): p. 631-640.

52.   Kuhn, B.K.a.H., 3DPRINTING AND INFILTRA TION OF TOOL STEELS, in Advances in Powder Metallurgy & Particulate Materials. 2005.

53.   Kernan, B.D., et al., Homogeneous steel infiltration. Metallurgical and Materials Transactions A: Physical Metallurgy and Materials Science, 2005. 36(10): p. 2815-2827. https://doi.org/10.1007/s11661-005-0278-x

54.   German, R.M., Liquid Phase Sintering. 1985, New York: Plenum Press. https://doi.org/10.1007/978-1-4899-3599-1

55.   Zhu, H.H., L. Lu, and J.Y.H. Fuh, Study on Shrinkage Behaviour of Direct Laser Sintering Metallic Powder. Proceedings of the Institution of Mechanical

Engineers, Part B: Journal of Engineering Manufacture, 2006. 220(2): p. 183-190. https://doi.org/10.1243/095440505X32995

56. Lenel, F.V., et al., Some observations on the shrinkage behaviour of copper compacts and of loose powder aggregates. Powder Metallurgy, 1961. 4(8): p. 25-36. https://doi.org/10.1179/pom.1961.4.8.002

57. Liu, Z.Y., et al., Sintering activation energy of powder injection molded 316L stainless steel. Scripta Materialia, 2001. 44(7): p. 1131-1137. https://doi.org/10.1016/S1359-6462(01)00664-9

58. Park, D.Y., et al., Effects of Particle Sizes on Sintering Behavior of 316L Stainless Steel Powder. Metallurgical and Materials Transactions A, 2013. 44(3): p. 1508-1518. https://doi.org/10.1007/s11661-012-1477-x

59. Crane, N.B., et al., Improving accuracy of powder-based SFF processes by metal deposition from a nanoparticle dispersion. Rapid Prototyping Journal, 2006. 12(5): p. 266-274. https://doi.org/10.1108/13552540610707022

60. Sachs, E., et al. Low shrinkage metal skeletons by three dimensional printing. in Proceedings of Solid Freeform Fabrication Symposium. 1999.

61. Sachs, E.M., et al., Metal and ceramic containing parts produced from powder using binders derived from salt. 2003.

62. Yoo, H.J., Reactive binders for metal parts produced by Three Dimensional Printing, in Dept. of Mechanical Engineering. 1997, Massachusetts Institute of Technology.

63. Hong, S.B., et al., Corrosion behavior of advanced titanium-based alloys made by three-dimensional printing (3DPTM) for biomedical applications. Corrosion Science, 2001. 43(9): p. 1781-1791. https://doi.org/10.1016/S0010-938X(00)00181-5

64. Hutchings, I.M. and G.D. Martin, Introduction to Inkjet Printing for Manufacturing, in Inkjet Technology for Digital Fabrication. 2012, John Wiley & Sons, Ltd. p. 1-20. https://doi.org/10.1002/9781118452943

65. Islam, M.N., H. Gomer, and S. Sacks, Comparison of dimensional accuracies of stereolithography and powder binder printing. International Journal of Advanced Manufacturing Technology, 2017. 88(9-12): p. 3077-3087. https://doi.org/10.1007/s00170-016-8988-3

66.   Ho, C.M.B., S.H. Ng, and Y.-J. Yoon, A review on 3D printed bioimplants. International Journal of Precision Engineering and Manufacturing, 2015. 16(5): p. 1035-1046. https://doi.org/10.1007/s12541-015-0134-x

67.   Chen, H. and Y.F. Zhao, Process parameters optimization for improving surface quality and manufacturing accuracy of binder jetting additive manufacturing process. Rapid Prototyping Journal, 2016. 22(3): p. 527-538. https://doi.org/10.1108/RPJ-11-2014-0149

68.   Onuh, S.O. and K.K.B. Hon, Optimising build parameters for improved surface finish in stereolithography. International Journal of Machine Tools and Manufacture, 1998. 38(4): p. 329-342. https://doi.org/10.1016/S0890-6955(97)00068-0

69.   Lanzetta, M. and E. Sachs, Improved surface finish in 3D printing using bimodal powder distribution. Rapid Prototyping Journal, 2003. 9(3): p. 157-166. https://doi.org/10.1108/13552540310477463

70.   Ohman, C., et al., Compressive behaviour of child and adult cortical bone. Bone, 2011. 49(4): p. 769-776. https://doi.org/10.1016/j.bone.2011.06.035

71.   Bandyopadhyay, A., et al., Influence of porosity on mechanical properties and in vivo response of Ti6Al4V implants. Acta Biomaterialia, 2010. 6(4): p. 1640-1648. https://doi.org/10.1016/j.actbio.2009.11.011

72.   Shen, H. and L.C. Brinson, A numerical investigation of porous titanium as orthopedic implant material. Mechanics of Materials, 2011. 43(8): p. 420-430. https://doi.org/10.1016/j.mechmat.2011.06.002

73.   Goriainov, V., et al., Bone and metal: An orthopaedic perspective on osseointegration of metals. Acta Biomaterialia, 2014. 10(10): p. 4043-4057. https://doi.org/10.1016/j.actbio.2014.06.004

74.   Matassi, F., et al., Porous metal for orthopedics implants. Clinical Cases in Mineral and Bone Metabolism, 2013. 10(2): p. 111-115.

75.   Curodeau, A., E. Sachs, and S. Caldarise, Design and fabrication of cast orthopedic implants with freeform surface textures from 3-D printed ceramic shell. Journal of Biomedical Materials Research, 2000. 53(5): p. 525-535. https://doi.org/10.1002/1097-4636(200009)53:5%3C525::AID-JBM12%3E3.0.CO;2-1

76. Will, J., et al., Porous ceramic bone scaffolds for vascularized bone tissue regeneration. Journal of Materials Science: Materials in Medicine, 2008. 19(8): p. 2781-2790. https://doi.org/10.1007/s10856-007-3346-5

77. Karageorgiou, V. and D. Kaplan, Porosity of 3D biomaterial scaffolds and osteogenesis. Biomaterials, 2005. 26(27): p. 5474-5491. https://doi.org/10.1016/j.biomaterials.2005.02.002

78. Gibson, L.J. and M.F. Ashby, The Mechanics of Three-Dimensional Cellular Materials. Proceedings of the Royal Society of London. Series A, Mathematical and Physical Sciences, 1982. 382(1782): p. 43-59. https://doi.org/10.1098/rspa.1982.0088

79. Wen, C.E., et al., Processing and mechanical properties of autogenous titanium implant materials. Journal of Materials Science: Materials in Medicine, 2002. 13(4): p. 397-401. https://doi.org/10.1023/A:1014344819558

80. Pasteris, J.D., B. Wopenka, and E. Valsami-Jones, Bone and tooth mineralization: Why apatite? Elements, 2008. 4(2): p. 97-104. https://doi.org/10.2113/GSELEMENTS.4.2.97

81. Shrestha, S. and G. Manogharan, Optimization of Binder Jetting Using Taguchi Method. JOM, 2017. 69(3): p. 491-497. https://doi.org/10.1007/s11837-016-2231-4

82. Gupta, M. and G.K. Meenashisundaram, Selection of Alloying Elements and Reinforcements Based on Toxicity and Mechanical Properties, in Insight into Designing Biocompatible Magnesium Alloys and Composites: Processing, Mechanical and Corrosion Characteristics. 2015, Springer Singapore: Singapore. p. 35-67. https://doi.org/10.1007/978-981-287-372-9_3

CHAPTER 6

Industry Frontiers and Applications

Abstract

This chapter aims to illustrate the window of opportunity for inkjet 3DP of metals for engineering applications. To do so, comparisons are made between several metal AM systems. Then, an overview of two leading companies that are actively working in metal AM using inkjet 3DP is provided. Finally, several case studies in which inkjet 3DP is successfully deployed are presented.

Keywords

Inkjet 3DP, ExOne, Höganäs, Case Study, Metal AM

6.1 Comparison of Various Metal AM Techniques

Comparisons between the capabilities of the most prevalent metal AM techniques are made in Table 6.1 in terms of machine specifications, geometric limitations, and commercially available materials. As can be seen from the table, the capability of inkjet 3DP is comparable with major competitors using laser or an electron beam source. It should be pointed out that although the build speed of inkjet 3DP is much faster in comparison with other techniques, the need of post-printing processes may compromise the overall lead time.

Table 6.1 Capabilities of different metal AM techniques.

Metal AM process	Laser melting	Electron beam melting	Laser powder deposition	Inkjet 3DP	Ref
Technology	Powder bed	Powder bed	Material deposition (directed energy deposition)	Powder bed	
An example of machine manufacturers	EOS	Arcam AB	Optomec	ExOne	
The largest/latest machine by the manufacturer	EOS M 400-4	Arcam Q20plus	LENS 850-R	M-Print	[1-4]
Build environment	Argon or nitrogen	Vacuum (10^{-7} bar)	Hermetically sealed	Normal atmospheric	
Building volume (mm)	400 × 400 × 400	350 × 380 (O/H)	900 × 1500 × 900	800 × 500 × 400	[1-4]
Build speed	< 28.8 cm³/hr	60 cm³/hr	0.5 kg/hr	˙ 60 layers/hr (> 1200 cm³/hr)	[4, 5]
Min. layer thickness (μm)	20	50	250	100	[5-7]
Print resolution (μm)	-	-	25	63.5 (X-axis), 60 (Y-axis), 100 (Z-axis)	[3, 7]
Accuracy (mm)	± 0.050	± 0.040	± 0.25	-	[3, 5]
Surface finish (microns Ra)	9-50	10-60	On the sides, 12- 25	> 13	[5, 6, 8]
Materials according to the manufacturer (number of alloys)	Maraging steel (1), stainless steel (5), nickel based alloys (3), cobalt chrome (3), Titanium (3), aluminum (1)	Titanium (3), cobalt chrome (1), nickel based alloys (1)	Steel (9), nickel based alloys (5), titanium (5), aluminum (5)	Steel (3), steels infiltrated with bronze (3), cobalt chrome (1), nickel based alloys (2), tungsten (1)	[6, 9-11]

* The digit in parenthesis represents the number of alloys available for each material.

6.2 Industry Frontiers

6.2.1 ExOne

The ExOne company, established in 2005, provides inkjet 3D printers and inkjet 3D printed products for industrial customers. ExOne uses the liquid BJ on powder bed method to directly fabricate industrial products made of metallic materials, including stainless steels, Inconel alloys, cobalt-chrome, and inkjet 3D printed iron based metals infiltrated with bronze. Furthermore, the company offers inkjet 3D printed ceramic based materials for indirect inkjet metal 3DP. Aqueous binders are used for direct inkjet 3DP of metallic parts up to the size of $800\times500\times400$ mm^3. The minimum level of details is around 1.5 mm and the accuracy of printed components ranges from ±0.13 to ±1.3 mm depending on the components' size [12].

Materials properties of directly inkjet 3D printed parts are tabulated in Table 6.2. As can be seen, the relative density of inkjet 3D printed IN 625 and 316L stainless steel after being sintered is more than 96%. Furthermore, mechanical properties of inkjet 3D printed IN 625 are comparable with HIPed fabricated counterpart, according to the published data sheet [13]. In addition, the relative density of inkjet 3D printed iron-based materials infiltrated with bronze, consisting of 40% bronze and 60% iron or stainless steels, is greater than 95% [14-16].

6.2.2 Höganäs AB

Höganäs deploys a proprietary technique based on inkjet 3DP technology to directly fabricate metallic components with the maximum longitudinal dimension of 50 mm. This unique inkjet technology offers a combination of high accuracy, good surface finish and tolerance that no other AM technique can match [17]. Geometrical capabilities of this technology include a maximum z-axis resolution of 35 μm, the minimum length of 300 μm, and the minimum wall thickness and hole size of 200 μm. Moreover, the relative density of sintered parts is about 96%, and surface quality is below Ra 10 μm [17]. Inkjet 3D printed components are currently restricted to stainless steels and their mechanical properties are presented in Table 6.2.

Table 6.2 Typical properties of available materials for industrial applications

Materials	Ultimate tensile strength (MPa)	Yield strength (MPa)		Modulus of elasticity (GPa)	Relative desntiy (%)	Hardn ess	Surface finish after sintering (μm Ra)	Ref
ExOne								
IN 625	676*,669**	290*,303**	51*, 41**	193*,200**	+97	84 HRb	-	[13]
316L stainless steel	517	214	34	165	+96	74 HRb	-	[18]
Iron infiltrated with bronze	407	255	17	131	+95	72 HRb	15	[16]
316 stainless steel infiltrated with bronze	407	234	8	148	+95	60 HRb	15	[15]
420 stainless steel infiltrated with bronze (Annealed)	496	427	7	147	+95	93 HRb	15	[14]
420 stainless steel infiltrated with bronze (Non-annealed)	682	455	2.3	147	+95	97 HRb	15	[14]
Höganäs AB								
316 stainless steel	520	180	50	-	-	55 HRb	-	[17]
17-4PH	1050	900	4	-	-	25 HRc	-	[17]

*Horizontal Direction (XY), **Vertical Direction (Z)

6.3 Case Studies

Table 6.3 represents several case studies in which metal inkjet 3DP has been used to advantageously solve real-world manufacturing limitations. For all cases, inkjet 3D printed components significantly reduce costs and time, along with additional benefits such as extended parts' life span or design freedom for some cases.

Table 6.3 Some cases of using direct inkjet 3DP to solve real world issues [12].

Part specification	Engineering sectors	Challenges	Conventional manufacturing details	Direct inkjet metal 3DP cost and lead time	Competitive advantages	Ref.
Impellers **Part Size: 10 cm diameter** 	Pump manufacturing	Decreasing lead time and cost of fabrication a new impeller for performance test	Casting Production time: 6 to 12 weeks Cost: $2,000-$3,500	Time: 3-4 weeks from Cost: $150-$600	Reducing prototype costs and lead time	[19]
Stator **Part Size: 7.5-12.5 cm** 	Pump manufacturing for downhole application	Extending part's life	Machining Cost: $400-$500	Cost: $75-$150 each	Extending part life as well as reducing costs	[20]

Strainer plates for decanter centrifuges **Part Size: 10×15 cm**	Mining	Extending part's life and difficulty in quality control	Multi-piece assemblies by welding	-	Extending part life as a result of optimized part design and homogeneity of the part, making quality control more readily.	[21]
Terminal End of prosthetic hand components **Part Size: 2.5-12.5 cm**	Biomedical	Looking for competitive manufacturing method	Investment casting or machining Production time: 2 to 8 weeks Cost: $250-$1,500	Time: 2-3 weeks Cost: $25-$150	Reducing costs and lead time along with capability of configuring geometrically sophisticated part designs	[22]
Bicycle lugs, brackets, dropouts, fork crowns **Part Size: 2.5-15 cm**	Arts and decorative	Personalized bicycles' parts	Investment casting Production time: 3 to 4 weeks Cost: $1,000 USD	Production Time: 4 days Cost: $425	Reducing costs and lead time together with freedom of creative aspect of design	[23]

Summary

This chapter presents the potential of inkjet 3DP technology as a metal AM technique. Inkjet 3DP is capable to compete with other well-known metal AM techniques in the light of its shorter build cycle together with its acceptable resolution and surface finish while operating exclusively under an atmospheric condition. To have a clear vision of metal 3DP, two examples of leading companies along with properties of their parts are outlined. As illustrated by the case studies, the future applications of metal inkjet 3DP to substitute conventionally manufactured components would be limitless and dependent on the imagination of working engineers.

References

1.   https://www.eos.info/systems_solutions/eos-m-400-4.

2.   www.arcam.com/wp-content/uploads/justaddbrochure-web.pdf.

3.   https://www.optomec.com/wp-content/uploads/2014/04/LENS_850-R_Datasheet_WEB_0816.pdf.

4.   http://www.exone.com/Systems/Production-Printers/M-Print.

5.   Karunakaran, K.P., et al., *Rapid manufacturing of metallic objects*. Rapid Prototyping Journal, 2012. **18**(4): p. 264-280.

6.   https://www.optomec.com/wp-content/uploads/2014/04/LENS_MATERIALS-WEB.pdf.

7.   http://www.exone.com/Systems/Research-Education-Printers/M-Flex.

8.   Chen, H. and Y.F. Zhao, Process parameters optimization for improving surface quality and manufacturing accuracy of binder jetting additive manufacturing process. Rapid Prototyping Journal, 2016. **22**(3): p. 527-538.

9.   www.arcam.com/technology/products/metal-powders/.

10. http://www.exone.com/Resources/Materials.

11. https://cdn1.scrvt.com/eos/caf948a1836f247e/de3dd9532a55/EOS_Materials_Broschure_Table_en.pdf

12. http://www.exone.com/.

13. www.exone.com/Portals/0/ResourceCenter/Materials/X1_MaterialData_AlloyIN625.pdf.

14. www.exone.com/Portals/0/ResourceCenter/Materials/X1_MaterialData_420SS.pdf.

15. www.exone.com/Portals/0/ResourceCenter/Materials/X1_MaterialData_316SS_rv4.pdf.

16. www.exone.com/Portals/0/ResourceCenter/Materials/X1_MaterialData_Iron.pdf.

17. https://www.hoganas.com/globalassets/media/sharepoint-documents/BrochuresanddatasheetsAllDocuments/Additivemanufacturingofsmallandcomplexmetalparts_July_2016_1666HOG.pdf.

18. www.exone.com/Portals/0/ResourceCenter/Materials/X1_MaterialData_316LSS_US.pdf.

19. www.exone.com/Portals/0/ResourceCenter/CaseStudies/X1_CaseStudies_All%2012.pdf.

20. www.exone.com/Portals/0/ResourceCenter/CaseStudies/X1_CaseStudies_All%209.pdf.

21. www.exone.com/Portals/0/ResourceCenter/CaseStudies/X1_CaseStudies_All%2013.pdf.

22. www.exone.com/Portals/0/ResourceCenter/CaseStudies/X1_CaseStudies_All%2010.pdf.

23. http://www.exone.com/Portals/0/ResourceCenter/CaseStudies/X1-CaseStudy-ideas2cycles.pdf.

Index

About the Authors

Dr Manoj Gupta is affiliated with the Mechanical Engineering Department at NUS, Singapore. To his credit are: (i) Disintegrated Melt Deposition technique and (ii) Hybrid Microwave Sintering technique. He has published over 450 peer reviewed journal papers and owns two US patents. A multiple award winner, he actively collaborate/visit Japan, France, Saudi Arabia, Qatar, China, USA and India.

Mojtaba Salehi holds both the Bachelor and Master of Science degrees in materials science and engineering. Currently, he is a PhD student in the Mechanical Engineering Department at NUS and doing research in inkjet 3D printing of metals.

Dr Sharon Nai Mui Ling is a Senior Scientist and Team Lead for 3D Additive Manufacturing at the Singapore Institute of Manufacturing Technology (SIMTech), Agency for Science, Technology and Research (A*STAR). Her research areas include (i) 3D additive manufacturing of metallic structures for structural and functional applications, (ii) design and development of new alloys and metal matrix composites for additive manufacturing, and (iii) metal powder atomization and powder processing. She has published 1 book, 2 book chapters and over 50 peer reviewed journal papers. She has also led several 3D additive manufacturing collaborative projects with industry and research partners.

Dr Saeed Maleksaeedi is a research scientist in the area of materials science and engineering at the Singapore Institute of Manufacturing Technology, Agency for Science, Technology and Research (A*STAR). He has been working in the area of additive manufacturing for more than 8 years and his research focus include (i) Powder processing of metals and ceramics, (ii) additive manufacturing processes including binder jet 3D printing and 3D stereolithography for biomedical applications, (iii) metal and polymer matrix composites and (iv) investment casting using 3D printed patters. Saeed has published 1 book chapter and over 30 peer reviewed journal papers and publications. He is also holding several patents and disclosures in the area of additive manufacturing. In the past several years, he has been working closely with industry to testify and validate additive manufacturing processes for fabrication of functional components.

www.ingramcontent.com/pod-product-compliance
Lightning Source LLC
Chambersburg PA
CBHW071650210326
41597CB00017B/2174